O
TA
705
.E5   Engineering geological
          maps.

O
TA
705
.E5   Engineering geo-
          logical maps.

Titles in this series:

Engineering geological maps

A guide to their preparation

Prepared by the Commission on
Engineering Geological Maps
of the International Association
of Engineering Geology

The Unesco Press   Paris 1976

The designations employed and the presentation of the material in this publication do not imply the expression of any opinion whatsoever on the part of the Unesco Secretariat concerning the legal status of any country or territory, or of its authorities, or concerning the delimitations of the frontiers of any country or territory.

Published by The Unesco Press
7 Place de Fontenoy, 75700 Paris
Printed by Imprimeries Réunies, Lausanne

ISBN 92-3-101243-6
French edition 92-3-201243-X

Preface

With a view to promoting global and regional synthesis of knowledge and the general advancement of geological science, and to providing the scientific basis for a better understanding of the world's mineral and land resources, Unesco has, for many years, been concerned with the preparation and publication of small-scale geological maps of various kinds. The present booklet is devoted to a particular aspect of this programme, namely engineering geological mapping.

The purpose of such maps is to show the distribution of specific geological phenomena and characteristics of rocks and soils affecting the engineering use of different terrains. The ever-growing demand for such maps has revealed the need for a standardization of principles, systems and methods. This is an urgent but at the same time a difficult problem which can best be solved through international co-operation.

The present guidebook, prepared for Unesco by the Commission on Engineering Geological Maps of the International Association of Engineering Geology, summarizes the views of an international commission of experts and incorporates experience from various countries in which engineering geological mapping is already undertaken at an advanced level. This book does not pretend to give detailed instructions for mapping, but rather to be a synthesis of present-day experience in this field. The views expressed are those of the authors and are not necessarily those of Unesco.

Unesco wishes to express its gratitude to all those who collaborated in the preparation of this text, and especially to Professor M. Arnould, President of the International Association of Engineering Geology, to Professor M. Matula of the Comenius University, Bratislava, President of the Commission on Engineering Geological Maps of the International Association of Engineering Geology, and to Professor W. R. Dearman of the University of Newcastle upon Tyne who kindly edited the text.

Contents

Introduction

1

Engineering geological mapping began to be developed with the first steps towards co-operation between geologists and engineers in the building of the larger engineering works such as tunnels, dams and railways. The first maps hardly differed from current stratigraphic-lithologic and tectonic-structural maps. Gradually, increasing demands by engineers for more and more quantitative geological data led to the appearance first in explanatory texts, then in enlarged map legends and finally on the actual geological maps, of more specific information on the technical aspects of geological phenomena and their engineering interpretation. Up to the present such interpretative maps are those in current use and may even be called engineering geological maps.

Developments in the theory and practice of mapping in engineering geology, however, have shown that such technical or interpretative or derived maps are not what true engineering geological maps should be.

The task of engineering geology is to provide engineers, planners and designers with such information as will help them to create engineering structures and to develop the country in the best possible harmony with the geological environment. Without harmony, every civil engineering work, and these are mainly dams, tunnels, highways, cities, industrial agglomerations and big open pit mines, interferes often to a considerable extent with the dynamic equilibrium of the geological environment. This may result in detrimental consequences which can affect not only the economy and durability but also the safety of the works.

The geological environment is a very complex multi-component dynamic system which cannot be studied in its entirety in connexion with construction works or other engineering activities. Using the method of model analysis a simplified picture has to be created of this system comprising only those components of the geologic environment which from the point of view of engineering geology are of a decisive significance: namely the distribution and properties of rocks and soils, groundwater, characteristics of the relief and present geodynamic processes. An engineering geological map, showing the distribution and spatial relationships of these basic components, can reflect the history as well as the dynamics of the development of engineering geological conditions; it enables a prognosis to be made of the influence of the environment on the engineering works, as well as to predict in which way the works will interfere with the environment. It is in a key position, and of immeasurable importance in the system of engineering geological information.

Naturally, such maps cannot replace a detailed investigation of a construction site, but will help both in the rational design of a site investigation and in the interpretation of the results.

Maps can be prepared for the most varied purposes: for example, in land-use planning for complex utilization and development of regions of varied character (including scarcely mastered areas with permafrost, semi-arid climate, seismic hazard) and of differing extents varying from whole state-territories to individual city districts. They can serve for certain specific purposes only, or may present a broader, multi-purpose view necessary for solving more general problems; they can serve as first steps in planning, as well as in the final stages of designing urban, industrial, transport, hydrotechnical or other constructions. Dependent upon purpose, maps may be of varied extent, scale and detail; they can have different contents and a varied choice of mapping attributes, as well as different aspects of their evaluation.

In all this variety and individuality, engineering geological maps—as with stratigraphical-lithological and tectonic maps—must embody certain conventions, a common classification as well as common principles, and a certain degree of standardization. The achievement of this is a considerable and very difficult task in international co-operation between engineering geologists.

At recent international congresses and symposia, and in the professional literature a wide ranging discussion has taken place on the problem of the principles of engineering geological mapping. These discussions embrace such topics as: what is an engineering geological map? What are its basic concepts and methodological background? How may various kinds of maps be classified according to their purpose, scale and content? What basic criteria should be used in the classification of rocks (as well as other phenomena) and territorial units on engineering geological maps? What are the problems of collection, interpretation and representation of engineering geological information? How can computers be used in the preparation of maps and what is the future of engineering geological mapping? What is the likelihood of international co-operation in standardization, terminology and in exchange of experience?

A confirmation of the importance of mapping and of the necessity of international co-operation towards its subsequent development was given when the General Assembly of the International Association of Engineering Geology (IAEG) established in 1968 as its first commission the

Working Group on Engineering Geological Mapping. The task of this commission was determined as: (a) to make clear the present situation in engineering geological mapping; (b) to analyse the various types of maps called engineering geological maps, or maps which are to serve for building, construction and land-use planning; (c) to outline the trends for the future development of engineering geological cartography and to present general recommendations on the information to be provided by a complex engineering geological map; and (d) to contribute to international exchange of information on this subject.

After the publication of reports on the present stage of engineering geological mapping in various parts of the world (*IAEG Bulletin*, No. 3, 4), the presentation of this brief guidebook is the first accomplishment of the commission in response to its stated aims.

The text is in four main chapters. A discussion of the principles of engineering geological mapping involves the definition and classification of engineering geological maps, the classification of rocks and soils, consideration of hydrogeological and geomorphological conditions, and the evaluation of geodynamic phenomena. This is followed by a description of the techniques which may be adopted for acquiring and interpreting data. After a brief reference to the usual methods of geological mapping, the special requirements and techniques of engineering geological mapping are dealt with. Finally the question of the presentation of data on engineering geological maps and the layout of a descriptive memoir are considered.

This guidebook sets out to answer such questions as 'What is an engineering geological map?' 'How is an engineering map made?' and 'How is engineering geological information presented on such a map?'. Future efforts of the IAEG Commission on Engineering Geological Maps will be devoted, among other things, to the techniques of engineering geological mapping and the amplification of the views set out in this guidebook. A very important part of future work will be an attempt at international co-operation in the selection and adoption of an agreed list of standard symbols for use on engineering geological maps.

The commission members realize that this guidebook is not an exhaustive treatment of the subject, and hope that more complete versions may be prepared in the future. Any comments and suggestions that would be of assistance in preparing future guides to engineering geological mapping would be greatly appreciated.

Members of the IAEG commission who have taken part in the preparation of this guide are:

Professor Milan Matula (Chairman), Department of Engineering Geology and Hydrogeology, Comenius University, Gottwaldo nam. 2, Bratislava (Czechoslovakia).

Professor W. R. Dearman (Editor), Department of Geology, Engineering Geology Unit, University of Newcastle-upon-Tyne (United Kingdom).

Professor G. A. Golodkovskaja, Geological Faculty, Moskovskij Universitet, Moskva 117234 (U.S.S.R.).

Professor Ing. M. Janjic, 1100 Beograd, Tolstojeva 5 (Yugoslavia).

Dr A. Pahl, Bundesanstalt für Geowissenschaften und Rohstoffe, 3 Hannover, Postfach 230153 (Federal Republic of Germany).

A. Peter, Service Géologique Régional Massif Central, 22 Avenue de Lempdes, 63800 Cournon d'Auvergne (France).

Mrs Dorothy H. Radbruch-Hall (Secretary), United States Geological Survey, 345 Middlefield Road, Menlo Park, CA 94025 (United States of America).

Principles

2

2.1 Introduction

The purpose of engineering geology is to provide basic information for the planning of land-use and for the planning, design, construction and maintenance of civil engineering works. Such information is needed to assess the feasibility of the proposed land-use or engineering undertaking, and for the latter to assist in the selection of the most appropriate type and method of construction, to ensure the stability of a structure in its natural setting, and to aid the performance of necessary maintenance. Engineering geological research and mapping are therefore mainly directed towards understanding the interrelationships between the geological environment and the engineering situation; the nature and relationships of the individual geological components; the active geodynamic processes and the prognosis of processes likely to result from the changes being made.

For each situation this implies a unique dynamic geological system of interrelated and interdependent phenomena and processes which can only be very incompletely understood and represented. The principal factors creating the engineering geological conditions of an individual site or area are the rocks and soils, water, geomorphological conditions and geodynamic processes.

A map provides the best impression of a geological environment, including the character and variety of engineering geological conditions, their individual components and their interrelationships. But it is a simplified model of the facts and the complexity of various dynamic geological factors can never be entirely represented. The degree of simplification depends mainly on the purpose and scale of the map, the relative importance of specific engineering geological factors or relationships, the accuracy of the information and on the techniques of representation used.

An engineering geological map should fulfil the following requirements:

1. It should portray the objective information necessary to evaluate the engineering geological features involved in regional planning, in the selection of both a site and the most suitable method of construction, and in mining.
2. It should make it possible to foresee the changes in the geological situation likely to be brought about by a proposed undertaking and to suggest any necessary preventive measures.
3. It should present information in such a way that it is easily understood by professional users who may not be geologists.

Engineering geological maps should be based on geological, hydrogeological and geomorphological maps, but must present and evaluate the basic facts provided by these maps in terms of engineering geology.

2.2 Definition of an engineering geological map

An engineering geological map is a type of geological map which provides a generalized representation of all those components of a geological environment of significance in land-use planning, and in design, construction and maintenance as applied to civil and mining engineering.

Geological features represented on engineering geological maps are:

1. The character of the rocks and soils, including their distribution, stratigraphical and structural arrangement, age, genesis, lithology, physical state, and their physical and mechanical properties.
2. Hydrogeological conditions, including the distribution of water-bearing soils and rocks, zones of saturated open discontinuities, depth to water table and its range of fluctuation, regions of confined water and piezometric levels, storage coefficients, direction of flow; springs, rivers, lakes and the limits and occurrence interval of flooding; pH, salinity, corrosiveness.
3. Geomorphological conditions, including surface topography and important elements of the landscape.
4. Geodynamic phenomena, including erosion and deposition, aeolian phenomena, permafrost, slope movements, formation of karstic conditions, suffusion, subsidence, volume changes in soil, data on seismic phenomena including active faults, current regional tectonic movements, and volcanic activity.

Engineering geological maps should include interpretative cross-sections and an explanatory text and legend. They may also include documentation data which have been collected for the preparation of the map. More than one map sheet may be required to show all this information.

2.3 Classification of engineering geological maps

Engineering geological maps may be classified according to purpose, content and scale.

2.3.1   According to purpose, they may be:
2.3.1.1   Special purpose, providing information either on one specific aspect of engineering geology, or for one specific purpose.
2.3.1.2   Multipurpose, providing information covering many aspects of engineering geology for a variety of planning and engineering purposes.

2.3.2   According to content, they may be:
2.3.2.1   Analytical maps, giving details of, or evaluating individual components of the geological environment. Their content is, as a rule, expressed in the title, for example, map of weathering grades, jointing map, seismic hazard map.
2.3.2.2   Comprehensive maps. These are of two kinds—they may be maps of engineering geological conditions depicting all the principal components of the engineering geological environment; on the other hand they may be maps of engineering geological zoning, evaluating and classifying individual territorial units on the basis of the uniformity of their engineering geological conditions. These two types may be combined on small-scale maps.
2.3.2.3   Auxiliary maps. These present factual data and are, for example, documentation maps, structural contour maps, isopachyte maps.
2.3.2.4   Complementary maps. These include geological, tectonic, geomorphological, pedological, geophysical and hydrogeological maps. They are maps of basic data which are sometimes included with a set of engineering geological maps.

2.3.3   According to scale, they may be:
2.3.3.1   Large-scale: 1 : 10 000 and greater.
2.3.3.2   Medium-scale: less than 1 : 10 000 and greater than 1 : 100 000.
2.3.3.3   Small-scale: 1 : 100 000 and less.

2.4 Principles of classification of rocks and soils for engineering geological mapping

The boundaries of rock and soil units shown on engineering geological maps of various scales should delimit rock and soil units which are characterized by a certain degree of homogeneity in basic engineering geological properties.

The main problem in engineering geological mapping is the selection of those geological features of rocks and soils which are closely related to physical properties, such as strength, deformability, durability, permeability, which are important in engineering geology. This is because, at present, we lack regional data on the variability of engineering properties of rocks and soils. Neither have suitable methods and techniques been developed for determining them in sufficient quantity, over large areas, quantitatively, quickly and cheaply. It is for this reason that we have to use those geological properties which best indicate physical or engineering geo-

logical characteristics. These are: (a) mineralogical composition closely related to specific gravity, Atterberg limits and plasticity index; (b) textural and structural characteristics, such as particle size distribution, related to unit weight, porosity; (c) moisture content, saturation moisture content, consistency, degree of weathering and alteration, and jointing, related to the physical state of soils and rocks and indicating strength properties, deformation characteristics, permeability and durability.

Classification of rocks and soils on engineering geological maps should be based on the principle that the physical or engineering geological properties of a rock in its present state are dependent on the combined effects of mode of origin, subsequent diagenetic, metamorphic and tectonic history, and on weathering processes. This principle of classification makes it possible not only to determine the reasons for the lithological and physical characteristic of soils and rocks, but also for their spatial distribution. This is a basic principle of engineering geological mapping as of other geological mapping, and implies not only the classification of individual rock samples but also the use of many individual rock samples, field observations and measurements to delineate uniform and continuous rock units.

The following classification,[1] based on lithology and mode of origin, is suggested: (a) engineering geological type (ET); (b) lithological type (LT); (c) lithological complex (LC); (d) lithological suite (LS). There will be different degrees of homogeneity for each unit.

The engineering geological type has the highest degree of physical homogeneity. It should be uniform in lithological character and physical state. These units can be characterized by statistically determined values derived from individual determinations of physical and mechanical properties and are generally shown only on large-scale maps.

A lithological type is homogeneous throughout in composition, texture and structure, but usually is not uniform in physical state. Reliable values of average mechanical properties cannot be given for the entire unit; usually only a general idea of engineering properties, with a range of values, can be presented. These units are used on large-scale, and where possible, on medium-scale maps.

A lithological complex comprises a set of genetically related lithological types developed under specific palaeogeographical and geotectonic conditions. Within a lithological complex the spatial arrangement of lithological types is uniform and distinctive for that complex, but a lithological complex is not necessarily uniform in either lithological character or physical state. In consequence, it is not possible to define the physical and mechanical properties of the whole lithological complex, but only to give data on the individual lithological types comprising the complex and to indicate the general behaviour of the whole lithological complex. The lithological complex is used as a mapping unit on medium-scale and some small-scale maps.

---

1. The classification adopted for engineering geological rock and soil units may be compared to the unit terms used in lithostratigraphical classification (Hedberg, 1972). The conventional hierarchy of lithostratigraphical terms is as follows:
   Bed = named or unnamed distinctive individual layer;
   Member = named or unnamed lithological entity within a formation;
   Formation = fundamental unit of lithostratigraphy;
   Group = two or more formations.
   The unit terms are used in engineering geology without any stratigraphical implications and in fact cannot be used in that way, and it is for this reason that the conventional lithostratigraphical terms were not used and a new set of terms were adopted specially for engineering geological use. There is, however, an approximate general equivalence of terms; for example engineering geological type = bed; lithological type = member; lithological complex = formation; lithological suite = group.

The lithological suite comprises many lithological complexes that developed under generally similar palaeogeographical and tectonic conditions. It has certain common lithological characteristics throughout which impart a general unity to the suite and serve to distinguish it from other suites. Only very general engineering geological properties of a lithological suite can be defined. These units are only used on small-scale maps.

On engineering geological maps the distribution of mapping units as well as their stratigraphical and structural arrangements and age relationships are shown. The engineering geological properties of the map units should be described in an accompanying explanatory legend. These map units will be used on both multipurpose and special purpose comprehensive or analytical maps.

2.5 Hydrogeological conditions

Hydrogeological conditions affect land-use, planning, site selection and the cost, durability and even the safety of structures. Ground and surface waters play a prominent part in such geodynamic processes as weathering, slope movements, mechanical and chemical suffusion, the development of karstic conditions, volume changes by shrinking and swelling, and collapse in loessic soils. Rock and soil properties are often changed by groundwater. Groundwater may influence excavation and construction methods by flowing into excavations, by producing seepage forces and uplift pressures and by its corrosive action. Hydrogeological conditions may also affect underground waste disposal.

Natural groundwater and surface water régimes may be directly influenced by hydraulic structures and by extraction of groundwater, and indirectly by factors such as urbanization and deforestation which increase runoff, sediment load in streams and erosion, thereby influencing other processes such as slope movement and sedimentation.

One aim of engineering geology, facilitated by the provision of hydrogeological data on maps, is the prediction of undesirable changes in the hydrogeological régime and the recommendation of procedures to avoid them. In engineering geological mapping, therefore, the following important information on hydrogeological conditions should be evaluated and represented on maps: the distribution of surface and subsurface water; infiltration conditions; water content; direction and velocity of groundwater flow; springs and seepages from individual water-bearing horizons; depth to water table and its range of fluctuation; regions of confined water and piezometric levels; hydrochemical properties such as pH, salinity, corrosiveness; and presence of bacterial or other pollutants.

On small-scale maps hydrogeological information is represented by symbols and numbers. On medium-scale maps the water table may be represented by contours and its range of fluctuation indicated by numbers. In mountainous regions this is not possible and depth to water table and other features can only be shown by numbers. Both depths to confined water and piezometric levels can be shown by contours. On large-scale maps hydrogeological conditions are represented by isohypses, isobaths and isopiestic lines, with known fluctuations shown numerically.

2.6 Geomorphological conditions

Geomorphological mapping is helpful in explaining the recent history of development of the landscape such as the formation of valleys, terraces, slope configuration and the processes active in the landscape at the present time. It is an essential part of engineering geological mapping which can be carried out quickly and cheaply and is often a decisive factor in planning an engineering geological investigation.

Evaluation of geomorphological conditions in engineering geological mapping should be more than a simple description of surface topography. It should include an explanation of the relationship between surface conditions and the geological setting; the origin, development and age of individual geomorphological elements; the influence of geomorphological conditions on hydrology and geodynamic processes. Also very important in engineering geology is the prediction of impending development of geomorphological features such as the lateral erosion of river banks, movement of dunes, collapse in karst or undermined areas.

Surface topography is shown by contours on maps of all scales. Point symbols are used to indicate significant geomorphological elements on small-scale maps. On medium and large-scale maps the actual boundaries and details of geomorphological features can be mapped.

2.7 Evaluation of geodynamic phenomena

Geodynamic phenomena are those geological features of the environment resulting from geological processes active at the present time. Excluded are depositional or alteration processes as these are included in the description of rock and soils units. The geological features include those due to erosion and deposition, aeolian processes, slope movements, permafrost, formation of karstic conditions, suffusion, volume changes in soil, seismic and volcanic activity. All these features are important in engineering geological planning and construction. They can be shown on special-purpose or multipurpose maps, and on analytical or comprehensive maps. The amount of detail shown depends on the scale of the map. It is important to show not only the features but also the conditions favouring their development, their intensity and frequency of occurrence.

Excessive erosion commonly produces many steep-sided gullies and ravines on hillsides and in extreme cases badlands. Erosion removes material and steepens slopes along streams, reservoirs and natural shorelines. Erosion of hillsides not only damages agricultural land but also causes construction problems. It creates an irregular surface, increases sediment load in streams and thereby increases erosion, removes lateral support from parts of slopes thus increasing the possibility of slope movements. Sediment washed from hillsides may accumulate in culverts, storm drains, gutters and other drainage facilities, or contribute to the rapid silting up of reservoirs.

Favourable conditions for excessive erosion are soft rocks of low permeability, moderate to steep slopes, sparse vegetation and high rainfall concentrated in a short period of time. Contributory factors are overgrazing, overcultivation, deforestation and urban development.

The erosion features commonly shown on engineering geological maps are hillside gullies and ravines, and river banks and shorelines that are being actively eroded.

Aeolian processes are generally among the less damaging geodynamic processes, but may be troublesome to engineering structures in certain areas. Dunes that develop in sandy arid and semi-arid regions can move across and block transportation lines which then require constant maintenance. Overgrazing, overcultivation or deforestation can create dune fields in some sandy areas. Conversely, dunes can often be stabilized by planting. Dunes and similar features should be shown on engineering geological maps.

Slope movements take place under the influence of gravity and include creep, slide, flow and fall of all types of rock and soil.

Geological conditions favourable for the development of slope movements are varied. They include hard resistant rocks overlying softer ones, such as volcanic rock over clay, or sandstone beds with shale overlying shales with minor intercalations of sandstone, or relatively undisturbed beds over rocks highly sheared by faulting; rocks that are highly jointed, fractured or sheared; hard and soft rocks alternating in a slope; unconsolidated sediments overlying relatively impermeable bedrock; presence of groundwater. Slope movements are caused or triggered off by other natural processes or by the activities of man. The conditions suitable for slope movements, and the features that are the result of such movements can be shown on maps; the factors that trigger the movement often cannot be shown.

The factors that cause slope movement can be divided into those that reduce shear strength and those that increase shear stresses on the slope.

Permafrost, permanently frozen ground, is widespread in arctic and subarctic regions. Construction problems can be expected in permanently frozen fine-gained materials such as silt, particularly those containing ice lenses and wedges. Certain features indicating permafrost can be shown on maps; for example, polygonal ground, thaw lakes and subsidence due to thawing after interference by man.

In northern regions, seasonal freezing and thawing of the ground, particularly in fine-grained materials, can also cause problems such as frost heaving of piles or damage to highways.

Karst features result from the solution of rocks. Common karst features on the surface are sinkholes, blind valleys and dry valleys with steep walls; underground, cave systems are also common. In addition the bedrock surface is extremely irregular and is usually covered by soils of varying compressibility.

Suffusion is the washing out of fine particles from unconsolidated materials, in particular sands and gravels. It is a minor geodynamic process, but may give rise to considerable problems in the design of hydraulic structures. The most common suffusion feature that can be shown on a map is the location of upwelling water or suffusion springs.

Volume changes in shrinking and swelling soils can cause damage to structures. Areas of such soils should be shown on engineering geological maps.

Geodynamic seismic features are the result of seismic activity recent enough for the effects still to be visible as geomorphological forms. In addition, it is sometimes possible to show on engineering geological maps areas of continuing relative tectonic uplift and depression as determined by geodetic measurements, and inferred active faults determined from historical records or geological data such as the juxtaposition of recent and older deposits, or raised and tilted terraces and shorelines. Features associated with active faults include offset streams, terraces and man-made structures; scarps; sag ponds; shutterridges; lines of springs; linear trenches.

Volcanic activity may have associated seismic activity and current local uplift and depression, but it is the frequency and intensity of the activity and the nature, location and extent of the volcanic products that may be of more importance in engineering geology.

It should be the aim of engineering geology not only to show the extent and distribution of geodynamic features, but also wherever possible to indicate their age and degree of activity.

On small-scale maps point data on geodynamic features can be shown by symbols. On medium-scale maps areas of the occurrence of geodynamic features should be delineated and the boundaries of individual features should be shown where possible. The actual boundaries of individual geodynamic features, and where possible their internal structures, can be shown on large-scale maps.

2.8 Principles of engineering geological zoning

Comprehensive engineering geological maps may present information in terms of engineering geological zoning. These are individual areas on the map which are approximately homogeneous in terms of engineering geological conditions and the area covered by any particular map sheet may be subdivided into a number of distinctive zoning units.

The detail and degree of homogeneity of each engineering geological zoning unit will depend on the scale and purpose of the map. For example, on small-scale maps the criterion of zoning would be the general uniformity in the main elements comprising the geological environment, such as geotectonic structure or regional geomorphological features. On larger scale maps, zones are based on an evaluation of the uniformity of the structural arrangement and composition of rock and soil units, on hydrogeological conditions, and on geodynamic phenomena.

Engineering geological zoning can be undertaken either for a general purpose or for a special purpose. On a map of general purpose zoning the following taxonomic natural territorial units would be recognized:
1. Regions, based on the uniformity of individual geotectonic structural elements.
2. Areas, on the basis of the uniformity of individual regional geomorphological units.
3. Zones, on the basis of the lithological homogeneity and the structural arrangement of lithofacial complexes of rocks and soils.
4. Districts, in which hydrogeological conditions and geodynamic phenomena are uniform.

In this way, the characteristics of a territory can be defined by zoning which, in turn, can then be used to evaluate the complexity of engineering geological conditions in individual territorial units for land-use and engineering purposes.

A map of special purpose engineering geological zoning would be prepared with a particular type of engineering undertaking in mind, for example, highways, dams, tunnels. On such a map the zoning units would be based on the analysis of geological phenomena and on geotechnical parameters, and evaluated in terms of a particular engineering purpose.

2.9 General principles

The main principles of engineering geological mapping should be applicable to maps of all types and all scales. If this is done, it will be possible to compare maps prepared at the same scale and at a variety of scales. The basic difference between maps at different scales should only be in the amount of data presented, and in the way in which information is presented. For example, the scale of the map will determine whether a landslide is represented by a point symbol appropriate to small-scale maps. A generalized symbol representing the type of landslide and occupying the actual area covered by the landslide would be used on medium-scale maps, whereas at large scales all the details within the landslide area would have been mapped to scale.

On engineering geological maps, of all types and at all scales, the information provided should be presented in such a way that not only the true nature but also the engineering significance of the data can be understood and fully appreciated.

2.10 Reference

HEDBERG, H. D. (ed.). 1972. An international guide to stratigraphic classification, terminology, and usage. *Lethaia*, vol. 5, p. 297-323. (International Subcommission on Stratigraphic Classification, report no. 7B.)

Techniques for acquiring and interpreting data

3

3.1   Introduction

Engineering geological mapping has much in common with geological mapping as the purpose of both types of mapping is to present information about the geological environment. From the point of view of the civil engineer one of the shortcomings of conventional geological maps is that rocks of markedly different engineering properties may be grouped together as a single unit because they are of the same age and origin. However, the scope of engineering geological mapping is wider as in addition to lithostratigraphical and structural information other components have to be considered. These mainly include the description and qualification of significant physical and engineering properties of rocks and soils, of the thickness and areal extent of geological formations, of groundwater conditions and of geodynamic phenomena.

A geological map provides the fundamental basis for engineering geological mapping. However, to meet the additional requirements of engineering geological maps specific methods and techniques are employed for gathering and interpreting engineering geological information. A difficulty which is inherent in both geological and engineering geological mapping techniques is that changes in the character of rocks and soils are often gradational and can occur both horizontally and vertically.

3.2   General methods of geological mapping

Geological maps are usually prepared by adding geological data to existing topographical maps, to topographical maps made specially for the purpose or to vertical aerial photographs. The accumulated geological information is interpreted and a synthesis of geological conditions, involving drawing structural and stratigraphical boundaries between defined units, may be prepared either at the same or a smaller scale.

3.2.1   PREPARATION OF THE TOPOGRAPHICAL BASE MAP

Where a topographical base map does not exist or is on too small a scale to be used for field-work, a map will have to be made specially as a base for the geological map. A topographical map may be produced before the geological survey starts either by conventional methods of ground survey or from vertical aerial photographs. Alternatively the geologist may produce his own topographical map as the geological observations are made. There are several methods of doing this depending on the accuracy required, and methods including the pace-and-compass method, the hand-level method, the altimeter method and the plane table method are fully described in various textbooks. Terrestrial photogrammetry may be useful in surveying steep rock slopes and foundation conditions. Use of this photographic technique is described in the literature.

3.2.2   ACQUISITION OF GEOLOGICAL INFORMATION

The work entailed in assembling geological information for the preparation of an engineering geological map follows a set pattern comprising a number of stages. A preliminary step is the search for existing geological information on the area to be mapped, supplemented by the study of maps and any existing aerial photographs. This is followed by reconnaissance of the area in which the available evidence is assessed and new geological, geomorphological and geodynamic information is gathered; samples may be collected for initial laboratory study, and the general survey work may be supplemented by geophysical tests and some unsophisticated subsurface sampling using, for example, portable drills and augers. The final stage, which may either be general or devoted to the elucidation of conditions associated with a particular construction site, will involve detailed field mapping (3.3), a variety of field investigations (3.4.1–3) and laboratory and *in situ* tests (3.4.4).

3.3   Special requirements for engineering geological mapping

3.3.1   ENGINEERING GEOLOGICAL DESCRIPTION OF ROCKS AND SOILS

The classifications of rocks and soils used by geologists are not satisfactory for engineering purposes because significant properties are not included in, and cannot always be inferred from, the usual geological description. It is, therefore, recommended for engineering geological mapping practice to use simple rock names supplemented by selected descriptive terms. These terms should be applied to both the rock material and the rock mass and should include a description of colour, grain size, texture, structure, discontinuities within the mass, weathered state, alteration state, strength properties, permeability and other terms indicating special engineering characteristics.

Adequate description of a rock or soil mass may require additional information including the dip and strike, or the attitude, of structures and discontinuities, the surface character of bedding planes and other discontinuities, the variability of structures and discontinuities, the details of the weathering profile. Of particular importance is the estimation of the degree of isotropy and homogeneity of the rock mass.

All these characteristics should be described by using semi-quantitative descriptive terms which have been defined for use in different countries.

3.3.2 MAPPING OF ROCKS AND SOILS FOR ENGINEERING PURPOSES

From the results of an engineering geological survey, the engineering geologist should aim to produce a map on which units are defined by engineering properties. In general the boundaries of these units could be expected to follow lithological boundaries, but the engineering property boundaries might well bear no relation to either geological structure or to stratigraphical boundaries. An example would be where deep weathering has differentially affected various rock types. The boundaries of rock and soil units shown on engineering geological maps of various scales should therefore, as has been stated before (2.4), delimit rock and soil units which are characterized by a certain degree of homogeneity in basic physical properties. Selection of an appropriate method for drawing boundaries to mapping units in the field depends in the first instance on the purpose for which the mapping is being undertaken. In turn, purpose will dictate an appropriate scale and scale will define the basic taxonomic or mapping unit which may be the lithological suite, the lithological complex, the lithological type or the engineering geological type.

There are suitable methods for mapping the boundaries of each of these units, and these are:
1. Lithological suite. The interpretation of existing geological maps; reconnaissance mapping; photogeology.
2. Lithological complex. Areal mapping with facial analysis to group together genetically related lithological types.
3. Lithological type. Detailed areal mapping and petrographic investigation.
4. Engineering geological type. Detailed investigation of the physical state of the rock or soil mass within a mapped lithological type.

Methods used in characterizing each of the basic taxonomic or mapping units include:
1. Lithological suite. Evaluation of probable rock behaviour from a knowledge of the properties of known rock types.
2. Lithological complex. Geophysical investigations in the field. Systematic boring and sampling in the field. *In situ* testing. Laboratory or field-laboratory testing of physical and index properties. Petrographic investigation and the evaluation of rock behaviour from a knowledge of the properties of known rock types.
3. Lithological type. Detailed petrographic investigation. Geophysical testing in the field. Systematic determination of index properties in the laboratory. *In situ* and laboratory testing of mechanical and other rock properties.

4. Engineering geological type. *In situ* testing of mechanical and other rock properties. Systematic laboratory testing of physical and mechanical properties.

Basic requirements for both investigation methods and characterization methods in delineating mapping units are summarized in Figure 1 in which the application of the methods to maps at successively larger scales is illustrated.

3.3.3 MAPPING HYDROGEOLOGICAL CONDITIONS

The principal hydrogeological conditions which need to be recorded or monitored in engineering geological mapping are of two types. The first is concerned with surface information, such as springs, seepages, rivers, lakes.

The second relates to subsurface information obtained from existing boreholes and wells or from exploratory boreholes made for the purpose. Hydrogeological conditions should be quantified wherever possible.

Springs and seepages both permanent and intermittent should be mapped; stream flow should be recorded and the direction and flow of underground streams, for example in karstic areas, should be determined

Boreholes should be used to provide information on piezometric levels, coefficient of permeability, storage coefficient, and groundwater chemistry.

Water samples should be collected for laboratory analysis. Special attention should be paid to the determination of pH and carbon dioxide and sulphate content as factors causing corrosion of engineering works. Wherever possible reference should be made to hydrogeological maps and publications if they are available.

3.3.4 MAPPING THE RESULTS OF GEODYNAMIC PROCESSES

The method adopted for the mapping of geodynamic phenomena (2.7) depends on the scale of the map. It is important to describe not only the features but also the conditions favouring, and the factors causing, their development. It is important to determine not only the extent of the various phenomena, but also their frequency of occurrence, the severity and the degree of activity and the rate at which each process is going on. An attempt should also be made to predict the future development of the geodynamic phenomena. Wherever possible each geodynamic system should be evaluated quantitatively or semi-quantitatively.

At small scales individual phenomena may be mapped from aerial photographs and by using other methods of remote sensing, or by a reconnaissance survey. Quantitative evaluation may be possible by using the past records of existing maps, by studying aerial photographs taken at different times in the past, or from historical and other archival records.

At large scales, on the other hand, it is possible to map the full geomorphological results of the activity, either by detailed topographical survey or from aerial photographs.

Detailed surface mapping may be supplemented by using boreholes and geophysical methods. The rate of individual processes may be determined by direct measurement in the field over a period of time. If the appropriate maps, photographs and archival material are available this aspect is aided by examination of successive editions of large-scale maps, by aerial photographs taken at intervals of time, and by analysis of other records.

3.4 Special techniques for engineering geological mapping

3.4.1 PHOTOGEOLOGY

Photo-interpretation is an important aid to engineering geological studies as it provides a rapid, relatively cheap and precise method for the first appraisal of a large area. The scale adopted is usually 1:10,000 to 1:30,000. Although the method may sometimes reveal features which cannot be detected on the ground, it may also miss important subsurface information. It is essential that the results of a photogeological survey should be supplemented by observations on the ground at selected localities.

Difficulties arise in the discrimination of rock and soil types, but these may be overcome by the analysis of resultant landforms and by differences in tones of colour on the photographs. The structural elements of the terrain, such as bedding, faulting and jointing, may be more easily appreciated and mapped on stereo-pairs of vertical aerial photographs rather than on the ground. In the same way, natural ponds, seepages, springs, swallow holes, submarine springs and other hydrological and hydrogeological features may be mapped. Variations in depth to water table and of the weathered mantle may also be detectable.

Photo-interpretation can aid engineering geological studies in soil mapping, in slope stability, drainage and materials surveys, in groundwater studies, and in the selection of routes, and the sites for reservoirs and dams.

New forms of imaging such as radar, microwave and infra-red linescan are becoming available to the photogeologist.

Stereoscopic ground photography can be used to study engineering geological conditions in steep or inaccessible cliffs, and temporary exposures in engineering excavations. Photogrammetric techniques may be used to quantify the results.

It should be appreciated that photo-interpretation is highly skilled work and the best results are obtained by specialists working in close co-operation with the geologist.

3.4.2 GEOPHYSICAL METHODS

The techniques of particular use in engineering geological mapping are resistivity and seismic methods used both on the surface and in boreholes. Geophysical measurements in general provide two classes of information:
1. Values for certain physical properties of rocks and soils and their variation over the map area. Such properties may be correlatable with other properties of the rock mass, such as degree of weathering or jointing, which are of concern to the engineer.
2. Determination of the depth to boundaries between rocks and soils of differing physical properties or to the water table, and the position of vertical boundaries, for example faults, between rock types.

The importance of geophysical methods is that they provide a rapid means of indirect and non-destructive assessment of *in situ* conditions.

3.4.2.1 *Resistivity measurements*

The method is based on the measurement of the electrical resistivity of the ground which is dependent primarily on porosity, fracturing, degree of saturation and the salinity of the pore-water. For the method to work effectively there must be a good contrast between physical properties such as is provided, for example, by shatter zones of high porosity in igneous and metamorphic rocks, and by weathered rock overlying fresh igneous and metamorphic rocks.

A combination of two methods is in common use. Electrical boring is the determination of the variation in resistivity with depth below a selected point. It is used to determine depth to water table, bedrock and other changes in lithology. Electrical profiling depends upon the interpretation in engineering geological terms of variation in resistivity of the ground at a predetermined fixed depth at different positions along a traverse. A set of electrical profiles, akin to a pattern of boreholes to a standard depth, gives a map of variation in resistivity.

3.4.2.2 *Seismic measurements*

Density and modulus of deformation of the rock and soil mass determine the velocity of transmission of seismic waves through these media. Seismic velocities can be determined by standard refraction techniques from the surface, or by shot-firing at various depths in one borehole and recording times of arrival at corresponding depths in adjacent boreholes. The refraction technique is used to determine depths to different refracting horizons, for example rockhead or base of the weathering zone, and depends on there being an increase in velocity of transmission with depth. This is generally the case.

The method is useful in outlining areas of fractured or weathered rock, tracing marker horizons, and determining the depth to bedrock surfaces particularly beneath alluvium.

3.4.3 BORING AND SAMPLING TECHNIQUES

Boring may be undertaken to provide undisturbed or disturbed samples of rocks and soils, or to provide a hole for *in situ* testing and the installation of instruments in the ground. A variety of methods is available including augering, percussive boring and rotary core drilling.

Samples must, as far as possible, be truly representative of the ground conditions, and must neither be contaminated by material dislodged from higher in the hole, nor modified by the loss of some constituents.

For visual examination and the determination of index properties disturbed samples may be all that are required. On the other hand, the determination of the physical properties of soils and rocks demands undisturbed samples as far as possible representative of the materials in their natural condition. Undisturbed soil samples are usually retained in the sample tubes, with the exposed ends waxed and capped for transport to the laboratory without change in moisture content. Rock cores are usually extruded with minimum force into specially constructed core boxes; samples may be wrapped and waxed if required.

The frequency with which undisturbed samples are taken, as well as their diameter and length, are determined largely by requirements of the investigation in hand. It is conventional to take an undisturbed sample at each change in lithology, and this is recommended as a minimum requirement.

Both the pattern and spacing of boreholes need to be flexible to take account of the geological conditions, and only rarely will a predetermined grid of boreholes driven to

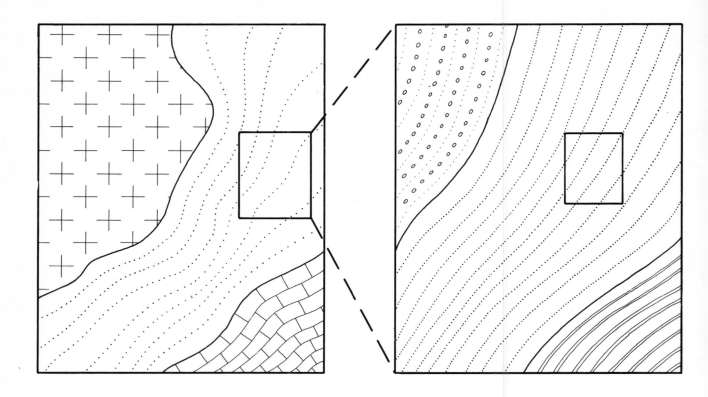

Map scale: < 1: 200,000

Taxonomic units shown on the map: Lithological suites

Attribute of homogeneity: Specific grouping of certain lithological complexes and their spatial arrangement

Methods of delimiting map units: Existing geological maps; reconnaissance mapping; aerial photogeology

Methods of characterizing map units: Evaluation of existing data

Map scale: 1: 10,000 to 1: 200,000

Taxonomic units shown on the map: Lithological complexes

Attribute of homogeneity: Specific grouping of certain lithological types and their spatial arrangement

Methods of delimiting map units: Areal mapping with facial analysis

Methods of characterizing map units: Boring and sampling; geophysical and petrographical investigation; laboratory determination of index properties

Description of map units:

 Granitic suite. Multiple granitic intrusion with remnants of schists comprising granite, granodiorite, amphibolite, paragneiss

 Lower Triassic clastic suite. Conglomerate, sandstone and mudstone complexes

 Middle Triassic calcareous suite. Limestone, dolomite and calcareous mudstone complexes (lithostratigraphical Cantal group)

Description of map units:

 Lower Triassic conglomerate complex. Littoral; conglomerates, sandstones and subordinate mudstones

 Lower Triassic sandstone complex. Flyschoid; calcareous sandstones, mudstones and subordinate siltstones (lithostratigraphical Omewa formation)

 Lower Triassic mudstone complex. Shallow neritic; mudstones with subordinate sandstones and siltstones

FIG. 1. The effect of scale on the basic requirement for the investigation and characterization of basic engineering geological mapping units.

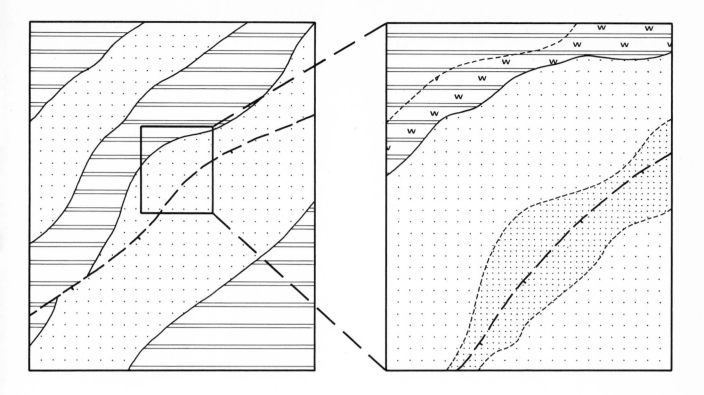

Map scale: 1: 5,000 to 1: 10,000

Taxonomic units shown on the map: Lithological types

Attribute of homogeneity: Mineralogical composition, texture and structure

Methods of delimiting map units: Petrographic investigation

Methods of characterizing map units: Boring and sampling; geophysical testing; limited *in situ* testing; systematic laboratory testing

Map scale: > 1: 5,000

Taxonomic units shown on the map: Engineering geological types

Attribute of homogeneity: Uniformity of physical state within each lithological type

Methods of delimiting map units: Investigation, for example, of degree of weathering, determination of discontinuity frequency and pattern, strength, consistency

Methods of characterizing map units: Determination of physical and mechanical properties

Description of map units:

 Light greyish-brown, fine to very fine-grained thinly bedded micaceous MUDSTONE

 Dark brown and yellowish-brown, coarse-grained, medium to thickly bedded calcareous SANDSTONE

Description of map units:

 Light greyish-brown, fine to very fine-grained, thinly bedded, closely jointed slightly weathered, micaceous MUDSTONE which slakes slowly on exposure, moderately weak

 Light greyish-brown, fine to very fine-grained, laminated, extremely closely jointed, moderately to highly weathered, micaceous MUDSTONE which crumbles in the fingers, very weak

 Dark brown, coarse-grained, medium to thickly bedded, with widely spaced joints, moderately to slightly weathered, calcareous SANDSTONE

 Yellowish-brown, coarse-grained, medium bedded, with closely to very closely spaced joints, sheared, highly weathered SANDSTONE

predetermined depths be entirely suitable for engineering geological investigations. Likewise, sampling should be dictated rather by the geological conditions than by a rigid system.

3.4.4 LABORATORY AND 'IN SITU' TESTING

3.4.4.1 *Laboratory tests*

Basic properties of rocks and soils may be determined by standardized laboratory tests.

Properties which are independent of moisture content include: particle size analysis; liquid and plastic limits; bulk density, both dry and saturated; mineral grain specific gravity; porosity (voids ratio); mineralogy and petrography.

Many of the tests to determine physical properties require undisturbed samples at their natural moisture content. They include: consistency; cohesion and angle of internal friction; compressibility; permeability; compressive strength; tensile strength; attrition value; compaction.

In engineering geological mapping, twenty-five to thirty samples are normally required for the statistical determination of the characteristics of each engineering geological type.

3.4.4.2 In situ *testing*

Sophisticated instrumental techniques are available for down-the-hole *in situ* determinations of, for example, the deformation characteristics of rocks and soils, the shear strength of soils, natural radio-activity, resistivity and spontaneous potential, and piezometric pressure.

Pumping-in and pumping-out tests may be carried out in existing boreholes to provide data on the hydrogeological characteristics of subsurface materials

The walls of boreholes may be observed, and the details recorded, by borehole cameras which may be linked to surface monitors.

Tests not requiring the use of a borehole include deep sounding tests. In these, static and dynamic penetrometers are used to determine the resistance of the ground to the penetration of a cone-shaped point. In the static penetration test the cone is jacked into the ground causing steady penetration; a free-falling drop-hammer is used in the dynamic test.

3.5 Analysis and interpretation of data

In carrying out an engineering geological survey of an area, information would have been gathered on all aspects of engineering geological conditions. The results obtained in the field and laboratory would include, for example, data on the distribution and properties of rocks and soils, on groundwater, and on both geomorphological conditions and geodynamic processes (2.1). This information would have been recorded directly on the field map sheet, in field notebooks, as borehole logs, and as the tabulated results of laboratory investigations.

Analysis of the data involves the selection and grouping of all the available information into those aspects which are considered of importance and absolutely necessary for the specific purposes for which the map is being made, and into

those which are not. The latter information is not processed further. At this stage an attempt should be made to assess the geological reliability of the data. This requires considerable geological experience and a sound appreciation of those general geological principles which may be applied to the area being mapped. Data that appear to be unreliable after this geological assessment should not be included in any further processing.

Having selected and grouped the information, the various groups may now be arranged in classes. This involves the use of various geological, engineering geological and engineering classifications adopted, for example, for the classification of rocks and soils according to their various properties, for the classification of groundwater and so on. The classification systems used may be those agreed upon internationally or those adopted by different countries. Classification is an important step in data processing; it enables considerable quantities of data to be arranged into sets each of which can be considered homogeneous.

A logical or a statistical approach may be used to assess the generalized qualitative or quantitative characteristics of each homogeneous set of data.

The final step in data processing is the synthesis of generalized information on the different individual components of the engineering geological conditions in order to determine and define individual territorial units that are characterized by a certain and specified degree of uniformity in their engineering geological conditions.

Beside these informal methods based on the skill and experience of the engineering geologist, computers are now beginning to be applied successfully to engineering geological mapping. There are three main areas of interest in the use of computers in this field: storage of coded data; statistical analysis of data and the correlation of a great number of variables; plotting or automatic drawing of computer produced maps.

3.6 References

BONDARICK, G. K. 1971. *Osnovy teoriyi ismenchivosti inzhenerno-geologicheskikh svoistv gornikh porod* [Principles of the theory of the variability of engineering-geological properties of rocks]. Moskva, Nedra.

BONDARICK, G. K.; KOMAROV, I. S.; FERRONSKIJ, V. 1967. *Polevie metody inzhenerno-geologischeskikh issledovanii* [Field methods of engineering-geological investigations]. Moskva, Nedra.

KOMAROV, I. S. 1972. *Nakoplenie i obrabotka informatsii pri inzhenerogeologischeskikh issledovaniach* [Acquisition and analysis of information in engineering-geological investigations]. Moskva, Nedra.

KRYNINE, D.P.; JUDD, W.R. 1957. *Principles of engineering geology and geotechnics.* McGraw-Hill Book Co. 730 p.

LAHEE, F. H. 1961. Field geology. 6th ed. New York, Harper.

LOW, J. W. 1957. *Geological field methods.* New York, Harper.

RENGERS, N. 1967. Terrestrial photogrammetry: a valuable tool for engineering geological purposes. *Rock mechanics and engineering geology,* vol. 5, p. 150-4.

TERZAGHI, K.; PECK, R. B. 1967. *Soil mechanics in engineering practice.* 2nd ed. New York, Wiley.

Presentation of data on engineering geological maps

4

4.1   Introduction

The content of engineering geological maps (2.3.3) and the amount of detail shown on engineering geological conditions (2.1) are determined by the purpose and the scale of the map. It is desirable to keep in mind the following important aim of engineering geological mapping, namely that the compilation of engineering geological maps should be based on the same general principles regardless of the fact that maps of different scales are designed to solve problems of different kinds. This would make it possible to compare directly maps drawn at the same scale for different areas, it would also facilitate the production of maps of medium to small scale, for example, from maps of large scale without the necessity of any radical changes.

There are many kinds of maps determined by the criteria of purpose, content and scale:

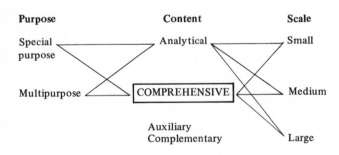

All combinations are possible; for example multipurpose maps may be prepared for a variety of engineering purposes covering many aspects of engineering geology; they may be analytical or comprehensive and may be prepared at all scales.

The most general or basic type of map is the multipurpose comprehensive map on which are presented and evaluated all the principal components of an engineering geological environment. This type of map would provide information pertinent to many purposes and needs. Techniques used in the presentation of data on a multipurpose comprehensive map are also of general application to other kinds of engineering geological maps. These techniques, considered according to scale, are discussed in detail below.

4.2   Multipurpose maps

Multipurpose maps may be analytical or comprehensive.

4.2.1   ANALYTICAL MULTIPURPOSE MAPS

The content of analytical maps is usually obvious from their title; for example an analytical map may be a map of intensity and pattern of jointing, or may show slope angles, or slope stability, or landslides. Thus an analytical multipurpose map gives both details of, and evaluates, an individual component of the geological environment for many purposes.

4.2.2   COMPREHENSIVE MULTIPURPOSE MAPS

On the other hand, comprehensive maps are of two basic kinds. They may show on one map sheet all the components of the engineering geological environment, or they may depict on one map sheet those areas which have been grouped for zoning purposes on the basis of the uniformity of their engineering geological conditions (zoning maps).

4.2.2.1   *Small-scale comprehensive multipurpose maps*

Small-scale maps of areas in which the engineering geology is well known can be compiled using available maps, literature and archival documents. In poorly investigated areas maps are prepared by photogeological interpretation and reconnaissance mapping.

Small-scale maps should show mainly the distribution and character of lithogenetic and lithofacial complexes of rocks and soils, depth to water table, corrosiveness of groundwater and the extent of active geological processes which are of importance in engineering geology.

Representation of rocks and soils is the main information contained on the map. It is recommended that the surficial deposits and the bedrock should be shown. The upper complex of surficial deposits may be shown by light colours selected according to a standard scheme[1] representing their lithogenetic character. Bedrock may be shown by darker colours or coloured patterns in which the colour indicates the genesis and the pattern the lithology of the rocks and soils. The age of rocks and soils may be represented by generally adopted geological symbols on the map or on the legend.

Hydrogeological data may be shown by isolines or numerically as point symbols in a generalized form for each complex of rocks and soils. Position of water tables, estimated quantity and details of groundwater chemistry may also be given.

Surface topography is shown by contours, and point symbols may be used to indicate significant geomorphological elements.

Selected point symbols may be used to show areas of active geodynamic processes, and seismic activity is shown by isoseisms.

On small-scale maps on which zoning is shown a general uniformity of the main components of the geological environment is the criterion adopted for discriminating individual zoning units. These could be regions on the basis of geotectonic elements, areas determined by macrogeomorphology, and possibly zones on the basis of uniformity in lithofacial character and structural arrangement.

4.2.2.2 *Medium-scale comprehensive multipurpose maps*

Medium-scale maps are prepared on the basis of field investigations and mapping supplemented by the use of existing archival material and any necessary complementary work. The same information should be shown on medium-scale maps as on small-scale maps but should be presented in greater detail. Lithological suites may be divided into lithological complexes and, if possible, into even smaller combinations of lithological types. Two or three uppermost rock and soil units may be shown on the map together with a lithological description of engineering geological terms. Rock and soil properties may be indicated in the legend.

In determining the depth to which engineering geological conditions should be shown, the main purpose as well as geological complexity will have to be taken into account.

As with small-scale maps, the rocks and soils at the surface may be shown by colour, and underlying complexes may be shown by distinguishing coloured patterns. If more than two rock and soil complexes are to be shown different methods of three-dimensional representation may be applied (4.5). Gradational values of thickness could be shown by different intensities of the colours used to distinguish various complexes, by variation in thickness of the coloured patterns, by isolines of thickness, or by numbers. Similar variations in colour or thickness of line in patterns could also be used for indicating different graduation in depth. Depths may also be shown by isolines or by numbers.

On medium-scale maps the water table may be represented by contours and its range of fluctuation indicated by numbers. In mountainous regions this is not possible and depth to water table and other features can only be shown by numbers. Both depths to confined water and piezometric levels can be shown by contours.

Data on groundwater chemistry should be shown on the map by symbols or numerically.

Surface topography is shown by contours and the actual boundaries and details of geomorphological features can be mapped. On medium-scale maps areas occupied by geodynamic features should be delineated and the boundaries of individual features should be shown where possible.

On medium-scale maps on which zoning is shown, zones are discriminated on the basis of the homogeneity and structural arrangement of rock and soil map units. Where possible, smaller areal units may be indicated in which either hydrogeological or geodynamic phenomena, or both, are uniform.

4.2.2.3 *Large-scale comprehensive multipurpose maps*

Large-scale maps are prepared by detailed field investigations and mapping, using all existing archival material, systematic subsurface exploratory and geophysical work and field and laboratory testing. It is necessary to determine the physical and mechanical properties of all rock and soil units represented on the map. Sufficient data may be available to permit statistical analysis of the results.

Lithological and engineering geological rock and soil types and their structure and spatial arrangement in depth may be shown by a combination of colours and coloured patterns (or simply in black and white). Two or more uppermost rock and soil units should be shown on the map together with a lithological description in engineering geological terms. Statistically determined rock and soil properties may be indicated in the legend.

In deciding the depth to which engineering geological conditions should be shown, the main purpose for which the map may serve will have to be taken into account as well as the complexity of the geology.

Different methods available for three-dimensional representation of thickness and depth conditions are dealt with in section 4.5.

On large-scale maps hydrogeological conditions may be represented at suitable intervals, say 1 m, by isohypses, isobaths and isopiestic lines, with known fluctuations shown numerically. The results of chemical analyses of groundwater may be shown numerically or by symbols. Other necessary data on water conservation areas and other aspects of hydrogeology may be shown.

Surface topograhy is shown by contours, and the actual boundaries and details of geomorphological features can be shown. The actual boundaries of individual geodynamic features, and where possible their internal structures, can be shown on large-scale maps.

Zoning on large-scale maps is based on the homogeneity and structural arrangement of the mapped rock and soil units, as well as on the uniformity of hydrogeological conditions and geodynamic phenomena.

4.3 Special purpose maps

Special purpose engineering geological maps are prepared for one specific purpose or provide information on one specific aspect of engineering geology. They may be analytical or comprehensive and are prepared at all scales.

Techniques used for representation of rock and soil units, hydrogeological conditions as well as geomorphology and geodynamic features are the same as those used for multipurpose maps (4.2.1 to 4.2.2.3).

---

1. There is at present no generally recognized international standard scheme of colours and symbols for use on engineering geological maps. However, there is an international legend for hydrogeological maps (Unesco, 1970).

4.3.1 ANALYTICAL SPECIAL PURPOSE MAPS

These maps represent individual components of the engineering geological conditions and evaluate them from the viewpoint of one specific purpose. For example, landslides may be evaluated in the context of urban development in which the engineering geological component is the landslide and the specific purpose for which the map is made is urban development; other examples involving landslides would be landslides and land reclamation, or landslides and the stability of reservoir slopes. To quote but a few examples, rocks and soils could be considered from the point of view of industrial use, of quarrying, of settlement characteristics of the ground and of water seepage from natural or artificial reservoirs; each of these combinations would be represented on a single analytical special purpose map.

4.3.2 COMPREHENSIVE SPECIAL PURPOSE MAPS

Comprehensive special purpose maps represent on one map sheet all the basic components of engineering geological conditions (2.1) and classify and evaluate them from the viewpoint of one specific purpose. The specific purpose could be urban development, underground construction or transportation routes as some examples.

Evaluation for different specific purposes could also be done on a map of engineering geological zoning by grouping territorial units on the basis of the uniformity of their engineering geological conditions. Such a zoning map could be a separate map sheet, or at small scales could be combined with the map of engineering geological conditions.

4.4 Interpretative geological maps

General purpose geological maps, even though they have not been made specifically for engineering purposes, contain a great deal of information of value to the engineer. An engineering geologist, or an engineer with a sound basic knowledge of geology, would be able to interpret such maps in engineering geological terms, but it should be borne in mind that the resultant interpretative map is not a true engineering geological map.

Interpretative geological maps may be provided by supplementation of geological maps with descriptive information in engineering geological terms by using additional legends on the published map sheet, by separately printed notes or by addition of information to archival copies of an existing geological map sheet.

4.5 Three-dimensional representation on maps

Because engineering structures have an influence on the subsurface, and often extend below the ground surface or are constructed below the surface, three-dimensional representation on engineering geological maps is desirable. Where surficial deposits overlie bedrock it is generally desirable to show the depth and character of the bedrock surface. On small-scale maps, three-dimensional information can be given only at points by symbols; in more detailed maps the bedrock surface and details of the surficial deposits should if possible be shown by any one of several methods: the stripe method; bedrock contours; point logs of borings; cross-sections (4.6); isometric diagrams; or isopachous lines. No method used should overcrowd the map to the point of illegibility. For showing on medium- and large-scale maps the depth and thickness of Quaternary deposits with only a few units between the ground surface and the bedrock surface, the stripe method is suggested as being particularly useful.

4.6 Engineering geological cross-sections

Engineering geological cross-sections are a necessary adjunct to all main types of engineering geological maps. The number and direction of cross-sections are chosen, taking into account geomorphology and geological structure, to illustrate the relationship between the components of the engineering geological conditions.

The horizontal scale of cross-sections should be equal to, or larger than, the scale of the map. Vertical scale is chosen so that it is possible to show the extent and character of the uppermost rock and soil units. The depth to which the cross-section is drawn should be directly related to the depth of available boreholes and other excavations.

All information presented on the map should also be shown on the cross-sections, for example hydrogeological conditions, engineering geological zoning, the results of geodynamic processes and engineering properties of all rock and soil units. The degree of detail on the cross-section will correspond with the detail shown on the map.

4.7 Documentation maps

A documentation map provides a record of sources of information used in compiling an engineering geological map. The documentation map should be drawn at the same scale as the associated engineering geological map, and if there is no danger of overcrowding the map with symbols the two maps may be combined on one sheet.

The information recorded on the documentation map includes, for example, symbols showing the position, depth and type of individual boreholes; observed or investigated exposures, pits, quarries and other excavations; wells and springs. Drill holes should be indicated in which pressuremetric tests and geophysical logging have been done; from which samples for laboratory tests were taken; which were tested by pumping water in or out, as well as those which were used for observation of the underground water régime. The location of sites used for engineering geological field experiments and geophysical experiments should also be indicated.

Information related to past investigations, photogeology and archives should also be outlined.

The legend on the documentation map sheet should explain all symbols used. The name of the organization which undertook each investigation, the date and the place where records are kept should be shown either on the map or in an accompanying memoir.

Standard symbols should be used where they are available.

4.8   Explanation and legend

The explanation or legend on the engineering geological map provides a guide to the symbols, colours and patterns used in producing the map. In addition, summary accounts may be given, for example, of rock and soil properties, hydrogeological conditions, geodynamic processes, and the evaluation of individual zoning units.

4.9   Reference

UNESCO. 1970. *International legend for hydrogeological maps/Légende internationale des cartes hydrogéologiques.* Paris, Unesco/IASH, IAH and London, Institute of Geological Sciences. 101 p.

Examples of engineering geological maps

5

5.1 Introduction

Maps chosen to illustrate the guidebook have not been drawn specially to illustrate the guidelines laid down in the text, but rather have been selected from the very wide range of engineering geological maps already available in published form. Published maps relate to actual geological situations which have been mapped in engineering geological terms, in a variety of ways. It was felt to be unwise to attempt to produce artificial maps to explain the principles of engineering geological mapping set out in the guide.

5.2 Examples of multipurpose engineering geological maps

5.2.1 MULTIPURPOSE ANALYTICAL MAPS

5.2.1.1 *Multipurpose, analytical, small-scale map*

Map of estimated abundance of landslides in the San Francisco Bay Region, California, at a scale of approximately 1:170,000.

Comment

The map was prepared as an experimental map to provide a first approximation of landslide abundance and potential in the area.

Different parts of the region are ranked on a scale from 1 to 6 according to the estimated abundance of landslides in them. Qualitative ranking is based on the area covered by landslides estimated on three factors: slope of the ground surface, rainfall, and rock and soil conditions. Inferred relations of landsliding to these factors, together with limited field knowledge of the actual abundance of landslides in some areas, have been used in the interpretation.

Method of map preparation

The region was divided into the six ranks of landslide abundance by progressively evaluating and ranking the high and low extremes. Rank 1 represents areas with a very small amount of landslides, and, at the other extreme, rank 6 represents areas that contain a maximum amount of landslides in the region. Areas ranked 5 have a lesser, but still very large amount of landslides, and areas with a limited amount of landslides, but larger than rank 1, are assigned to rank 2. The remaining, intermediate areas are assigned as appropriate to ranks 3 and 4.

Rank 1 is defined as areas that receive less than 10 in (255 mm) of mean annual precipitation or have slopes of less than 5° (determined from 1:500,000 scale topographic map with 500 ft (150 m) contour interval). These criteria indicate a very small amount of landslides, because landslides are rare in areas with less than 10 in (255 mm) of precipitation, and review of a limited number of reports and consultation with other geologists indicates that few landslides occur on slopes of less than 5° (Radbruch, 1970, p. 4–6). Because of the large contour interval on the map from which slopes were determined and the small scale of the final map, sea cliff areas and hilly areas with less than 500 ft (150 m) relief may locally contain abundant landslides, although shown on the map as rank 1.

Ranking of the remaining region is based largely on the distribution of earth materials. For this purpose the geologic units in the region that are shown on the 1:250,000 scale Jenkins edition of the *Geologic Map of California* were grouped into eight general classes of earth materials. The classes were selected to have as similar landslide characteristics as possible, using readily available literature on landslides in the region as a guide. Major differences in landslides characteristics within individual classes of earth materials resulting from varying topography or bedrock type or structure are used in the ranking where the needed information is available.

This example has been redrawn from a part of D. H. Radbruch and C. M. Wentworth, *Estimated Relative Abundance of Landslides in the San Francisco Bay Region, California*, 1971, United States Department of the Interior, Geological Survey (San Francisco Bay Region Environment and Resources Planning Study, Basic Data Contribution 11).

Legend

LANDSLIDE ABUNDANCE RANKS

1 — Least abundant

2

3

4

5

6 — Most abundant

Ranking is qualitative, based on estimates and extrapolation from available data. Specific safety or hazard for construction is not shown. Landslide distribution within individual map units may not be uniform: parts of the highest ranked units lack landslides, and parts of the lowest ranked units contain landslides. Hilly parts of unit 1 may contain abundant landslides

—— Approximate contact between map units

▬▬ Fault, approximately located, marking zone of possible sheared or shattered rock not represented in the ranking but susceptible to landsliding. Not shown where concealed beneath thick overlying deposits or water

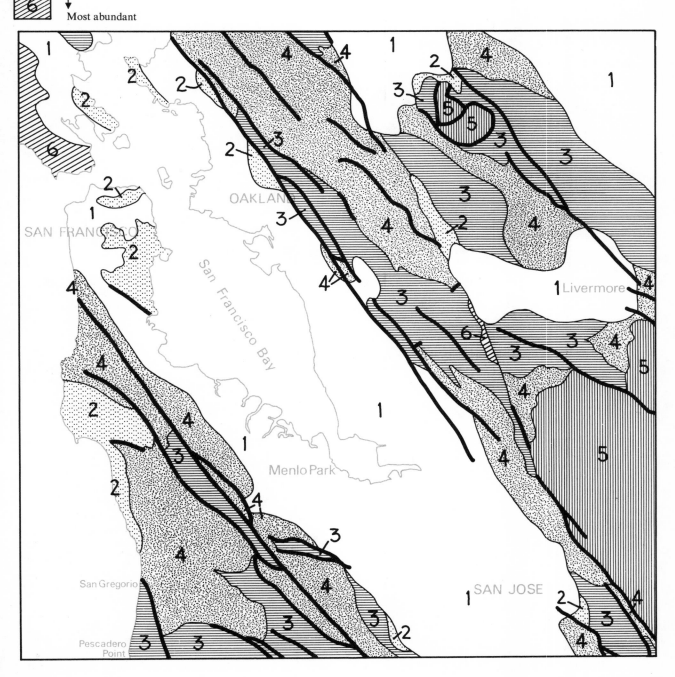

5.2.1.2   *Multipurpose, analytical, medium-scale map*

Map of landslide susceptibility, San Mateo County, California, at a scale of 1 : 54,500.

Comment

This map sheet provides a more detailed appraisal of landslide susceptibility than the smaller-scale map 5.2.1.1. A very full explanation is published on the map sheet, including notes on how to use the map, discussion of factors affecting landslide distribution and a slope intervals chart. The present map was compiled from the existing geological and landslide maps and specially produced slope map. Mapping procedures are discussed. A table gives the landslide failure record for the rock units cropping out in San Mateo County.

It is concluded that this slope stability map provides an easily read analysis of selected geological factors that bear directly on the problem of landsliding within the county.

The map has been redrawn from part of E. E. Brabb, E. H. Pampeyan and M. G. Bonilla, *Landslide Susceptibility in San Mateo County, California*, 1972, United States Department of the Interior, Geological Survey (Miscellaneous Field Studies Map MF-360) (San Francisco Bay Region Environment and Resources Planning Study, Basic Data Contribution 43).

Legend

Least

I — Areas least susceptible to landsliding. Very few small landslides have formed in these areas. Formation of large landslides is possible but unlikely, except during earthquakes. Slopes generally less than 15 per cent, but may include small areas of steep slopes that could have higher susceptibility. Includes some areas with 30 per cent to more than 70 per cent slopes that seem to be underlain by stable rock units. Additional slope stability problems; some of the areas may be more susceptible to landsliding if they are overlain by thick deposits of soil, slopewash or ravine fill. Rockfalls may also occur on steep slopes. Also includes areas along creeks, rivers, sloughs and lakes that may fail by landsliding during earthquakes. If area is adjacent to area with higher susceptibility, a landslide may encroach into the area, or the area may fail if a landslide undercuts it, such as the flat area adjacent to sea cliffs.

II — Low susceptibility to landsliding. Several small landslides have formed in these areas and some of these have caused extensive damage to homes and roads. A few large landslides may occur. Slopes vary from 5-15 per cent for unstable rock units to more than 70 per cent for rock units that seem to be stable. The statements about additional slope stability problems mentioned in I above apply in this category.

III — Moderate susceptibility to landsliding. Many small landslides have formed in these areas and several of these have caused extensive damage to homes and roads. Some large landslides likely. Slopes generally greater than 30 per cent but includes some slopes 15-30 per cent in areas underlain by unstable rock units. See I for additional slope stability problems.

IV — Moderately high susceptibility to landsliding. Slopes all greater than 30 per cent. These areas are mostly in undeveloped parts of the county. Several large landslides likely. See I for additional slope stability problems.

V — High susceptibility to landsliding. Slopes all greater than 30 per cent. Many large and small landslides may form. These areas are mostly in undeveloped parts of the county. See I for additional slope stability problems.

VI — Very high susceptibility to landsliding. Slopes all greater than 30 per cent. Development of many large and small landslides is likely. Slopes all greater than 30 per cent. The areas are mainly in undeveloped parts of the county. See I for additional slope stability problems.

Most

L — Highest susceptibility to landsliding. Consists of landslide and possible landslide deposits. No small landslide deposits are shown. Some of these areas may be relatively stable and suitable for development, whereas others are active and causing damage to roads, houses and other cultural features.

Definitions: Large landslide—more than 500 ft (150 m) in maximum dimension; small landslide—50 to 500 ft (15 to 150 m) in maximum dimension.

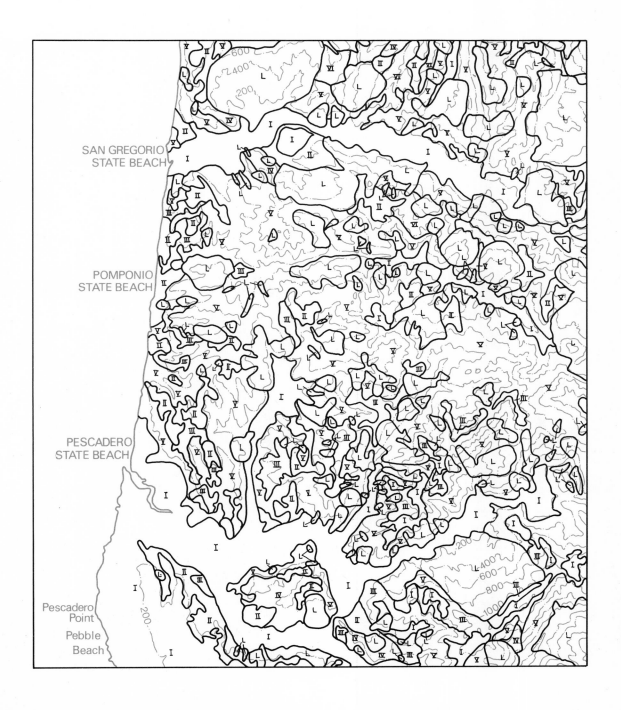

5.2.2   MULTIPURPOSE COMPREHENSIVE MAPS

5.2.2.1   *Multipurpose, comprehensive, small-scale map*

Engineering geological map of the Zvolen area (Czechoslovakia) at a scale of 1 : 200,000.

Legend

2 - 5   Orientation data on the depth of groundwater level (in metres).

Important springs (wells) of mineral water

Wider protection zones of groundwater utilized for spas

pH — CO_2, H — SO_4 Corrosiveness of groundwater

Landslides

Extensive gully erosion

Important faults: (a) verified; (b) assumed

Boundaries of lithological complexes: (a) at the surface; (b) under Quaternary deposits

Boundaries of engineering-geological areas

Boundaries of engineering-geological zones

Dm , Vh   Symbols for zones

Ce , Dg   Symbols for areas

Proluvial cones

Comment

The map was prepared for the purposes of regional and land-use planning.

Basic engineering geological data on rocks, groundwater and geodynamic phenomena are represented, and on the basis of their uniformity engineering geological areas and zones are delimited.

Rocks and soils are divided into lithological suites and complexes. Colours are used for the representation of lithological character and patterns for the lithology of Quaternary surficial materials (in ochre) and pre-Quaternary basement (in grey). Quaternary deposits are shown only where thicker than 3 m. The age of rocks is shown only in the legend by geological symbols.

Hydrogeological conditions are represented in a generalized form for each complex of rocks and soils by numerical symbols in blue colour. Point symbols in red are used to show areas of technically important geodynamic phenomena (e.g. landslides, gully erosion, karst phenomena).

The method of zoning is applied to delimit different types of engineering geological areas and zones. The discrimination of areas is based on the uniformity of individual regional geomorphological units, and zones are delimited on the basis of the general character and structural arrangement of lithological complexes.

The illustration is a small part of a map by R. Ondrasik and M. Matula (Czechoslovakia), at a scale of 1 : 200,000.

I. GENETIC-LITHOLOGICAL CLASSIFICATION					II. ENGINEERING CLASSIFICATION
Age	Lithological suites	Symbols	Patterns	Lithological complexes	A_1 = solid rocks, A_2 = semisolid rocks, B = gravelly soils, C = sandy soils, D = cohesive soils, E = unsuitable soils (classification according to ČSN Building Standard)
Quaternary	Surficial deposits	$_dQ_{3-4}$		Slopewash sediments	D = loams to stony loams (B = stony debris, talus)
		$_fQ_{3-4}$		Fluvial sediments on the (a) flood plain (b) terraces	B = sandy and loamy gravels, C = loamy sands, D = sandy loams (E = muddy soils)
		$_{pr}Q_{2-3}$		Sediments in proluvial cones	B = loamy gravels, D = sandy and stony loams
		$^tN_2-Q$		Chemical sediments	A_2 = travertine
Neogene	Molasse suite	$_lN_2$		Freshwater lacustrine and lacustrine fluvial complex	B = loamy gravels, C = loamy sands, D = sandy and silty clays
	Postero-genetic neovolcanites	β		Neovolcanic basalts	A_1 = basalt, A_2 = tuffs
		α		Neovolcanic andesites	A_1 = andesites, B_2 = tuffs (propylitized andesites)
		tu_α		Volcanic-lacustrine stratified andesitic tuffs and tuffites	A_2 = stratified tuffs, tuffites (A_1 = andesite), B = gravels (D = clays)
Mesozoic	Limestone-dolomite suite	ld_{T2}		Shallow-neritic calcareous complex	A_1 = limestones, dolomites
	Lower terrigenous suite	q_{T1}		Littoral quartzitic complex	A_1 = quartzite (conglomerate), A_2 = shales
Palaeozoic	Variscan granitoid intrusions	γ		Granitoids	A_1 = granodiorite to quartz diorite

(continued overleaf)

5.2.2.1 *continued*

III. ENGINEERING-GEOLOGICAL CHARACTERISTICS

For rock and soil material (in laboratory)

Symbols	Main lithological types	Dry bulk density g. cm^{-3}	Porosity %	Uniaxial compressive strength kgf. cm^{-2}.10^2	Indentation hardness (Šrejner) kgf.mm^{-2}	Modulus of deformation E$_o$ kgf. cm^{-2}.10^5	Durability
dQ$_{3-4}$	Sandy to clayey loam	1.53–1.79	35–58	0.0055	—	—	Soften, swell, not fr resistant
fQ$_{3-4}$	Sandy loam Sandy gravels	1.50–1.65 1.80–1.90	35–45	— —	— —	— —	Not frost resistant
prQ$_{2-3}$	Sandy loams and stony-loamy soils						
'N$_2$–Q	Travertine	1.60–1.95	1–10	1–3	150–190	—	Slightly soluble
$_l$N$_2$	Loamy gravels Sandy clays, clays	— 1.50–1.75	— 30–35	— —	— —	— —	Swell, soften
β	Basalts Basaltic tuffs	2.85–2.95 —	1.5–3.0 —	19–25 —	700–900 —	6–8 —	Very stable Slightly affected by
α	Andesites Block and agglomeratic stuffs	2.67–2.71 1.83–2.12	1.5–4.5 18–33	15–23 1.6–3	500–700 —	5–7 —	Stable Stable
tu$_α$	Tuffs Tuffites	1.30–1.70 1.68–1.85	30–40 29–37	1–2.5 1–1.15	50–80 —	0.4–0.8 —	Variable, not frost resistant, slake —
ld$_{T2}$	Limestones Dolomites	2.70–2.73 2.71–2.82	0.23–1.39 0.68–3.79	7–14 6–15	200–300 120–400	0.03–56 3–6	Stable
q$_{T1}$	Quartzites	2.53–2.58	3.0–4.0	10–14	380–420	3–3.4	Stable
γ	Granodiorites and quartz-diorites	2.62–2.70	1.1–2.7	12–17	270–600	0.4–0.6	Stable

ock and soil complexes (*in situ*)

of
ation

Thickness 1–9 m, fragmentary debris at the base, very low permeability, without permanent groundwater table, much gully erosion and landsliding used for brick materials, impervious materials for dams.

(a) thickness up to 3 m, permeable with permanent continuous groundwater table
(b) thickness up to 11 m, slightly permeable (10^{-4} to 10^{-3} m/s), without a permanent water table; suitable as fill material

Conical forms; coarse—alternating with fine—grain material

Dome-shaped bodies, stratified structure, open fissures, slightly permeable along fissures and cavities, slightly soluble in water; used for construction purposes

Cross-bedded sandy gravels and sands with alternating silty clays and clays; lateral transitions and wedging out. Sands and gravels are dense and slightly to moderately permeable; suitable for fills, locally may be suitable for concrete. Lenticles of cohesive soils are well consolidated, slightly permeable to effectively impermeable; used for brick making

Outliers and remnants of lava flows several metres to tens of metres thick, predominant prismatic jointing. Homogeneous, except marginally. Intense weathering to several metres, moderately permeable along fissures. Suitable for building stone and crushed aggregates. Irregular bodies of ashy to coarse-clastic tuffs, very heterogeneous, permeable along fissures and through pores. Suitable for light construction material

(a) irregular andesitic bodies several to 100 m thick, homogeneous except at the margins, predominantly prismatic jointing, frequent tectonic disturbance, sometimes propylitized. Fissure permeability. Suitable as a construction material
(b) tuffs tens to hundreds of metres thick, very heterogeneous tuffs and agglomerates, variably cemented, permeable through pores and along joints. Of limited use as a construction material

Tens to hundreds of metres thick irregular horizons of chaotic and stratified tuffs alternating with tuffites of varied grain size and degree of heterogeneity. Irregular jointing and frequent selective weathering. Permeable through pores and fissures. Of limited use as a construction material

Thick to massive bedded relatively homogeneous limestones and dolomites. Limestones are highly karstified and highly permeable. Used as a construction material; purer horizons are used for lime production
Massive dolomites with intercalations of shale and limestone are tectonically disturbed. Weathered to various depths to dolomite sand containing dolomite fragments. Locally used as fill material

Thickly bedded to massive quartzites with horizons of sericitic and chloritic shales, dynamometamorphosed to a varying degree. Weathered to a fine rubble. Permeable in tectonically disturbed zones. A suitable material for crushed aggregates

Massive, homogeneous, in places tectonically disturbed and dynamometamorphosed. Polyhedral jointing. Permeable along fissures and disturbed zones. A suitable construction material

(continued overleaf)

5.2.2.1 *continued*

AREAS	ZONES			
	Type	Geological conditions of the ground	Hydrogeological conditions	Present geodynamic processes
Ce Areas of the volcanic highlands — Young and variegated mountainous landscape, formed by the destruction of original stratovolcanic forms due to differencial neotectonic movements and erosion	**Dm** Zone of slopewash deposits on magmatic and metamorphic rocks	Loamy slopewash soils 2–7 m thick. In the substratum there are solid and semi-solid, slightly compressible volcanic and pyroclastic rocks	Very slightly water-bearing to practically dry	Intense slope erosion in esive furrows and gullies
Dg Areas of intramountain basins — Tectonically originated regional depressions, in which selective erosion and accumulation of soft Pliocene and Quaternary sediments conditioned moderate to flat forms of landscape/large river floodplains and terraces dominating	**Fu** Zone of floodplain alluvial deposits	Coarse-grained gravels, sandy gravels, sands and sandy to clayey loams. Large rivers may also have sapropelic fills in abandoned river branches and depressions. Thickness of alluvium 5–11 m	Gravel deposits are highly saturated. The groundwater level is at a depth of 5 m. Large areas often flood at high groundwater levels	Washing out of the bank causes small slope failure some places
	Nk Zone of pre-Quaternary alternating cohesive and uncohesive sediments	Moderately compacted Pliocene gravels, with silty-clayey bond, and weakly consolidated Pliocene clays of higher plasticity	On the gravelly Pliocene complex abundant small hillside springs may occur on slopes	Intensive slope erosion in form of deep furrows. Frequent small slides at cont with tuffites

Note: In this table the characteristics for only two selected types of areas and three types of zones are presented as an example.

NEERING-GEOLOGICAL CONDITIONS FOR CONSTRUCTION WORKS

vations and gs	Fill construction	Structural foundations	Roads	Building materials
tings and side-slope will mainly be vated in slopewash osits and the under-g weathered rocks; nd water issuing n the slope is easily ected and drained n the site	The construction of larger fills may lead to slope failure. Use of a combination of half fill-half cut would be advantageous	Foundation conditions are very good. The only serious problem is the question of ensuring slope stability during construction	When determining the line of roads and railways difficulties arise because of the effect of the gullies, and the need for good drainage to maintain stability	Brick clays
rule foundation vations have to be e below ground-er level. In deeper vations in gravels e influxes of water	Does not cause difficulties, but strongly compressible non-load-bearing sludgy and putrified muddy accumulations in oxbow lakes and the fringes of alluvial cones require attention	In foundation of buildings with cellars difficulties are caused by the high water table. Necessity of installing waterproofing (uplift pressure), in some cases precautions against corrosiveness. Fluctuating groundwater level and frequent flooding deteriorate the foundation materials	With regard to its very slight relief the zone is convenient. Difficulties occur in districts of marshes and oxbow lakes	Abundant stocks of high quality gravel and sand for the production of concrete. Abundant fill material
tings are excavated tly in moderate es. Eventual ndwater influxes n the slope are easily ected and drained n the site	The construction of bigger fills may lead to slope failure. Use of a combination of half fill-half cut of smaller dimensions would be advantageous	This zone is less suitable for heavy construction of larger settlements and industrial structures. It is mainly suitable for simpler structures	On account of clayey materials and the presence of deep scours, the zone is only conditionally suitable for transport constructions	Possible advantageous large gravel pits and clay pits. Gravels of Pliocene complex are, however, of lesser quality

5.2.2.2 *Multipurpose, comprehensive, medium-scale map*

Map of engineering geological conditions in part of the Zvolen Basin (Czechoslovakia) at a scale of 1:25,000 (Matula, 1969).

Legend

 Hydroisobaths of the max. groundwater table

 Springs of fresh and mineral water

 Waterlogged territory

 Corrosiveness of groundwater: pH degree, hardness of water (in German degrees); the full sector indicates the aggressivity of CO_2, and/or SO_4 according to Czechoslovak standards

Gully erosion

Formation of alluvial cones

Active landslides

 Older (potential) landslides

Edges of river terraces

Boundary of lithological complexes on the surface

a Boundary of complexes under the surface between:

(a) Quaternary deposits and (b) Quaternary deposits and pre-Quaternary basement; between pre-Quaternary complexes

b

⋋ 10 Strike and dip of strata

Important tectonic faults and their dip

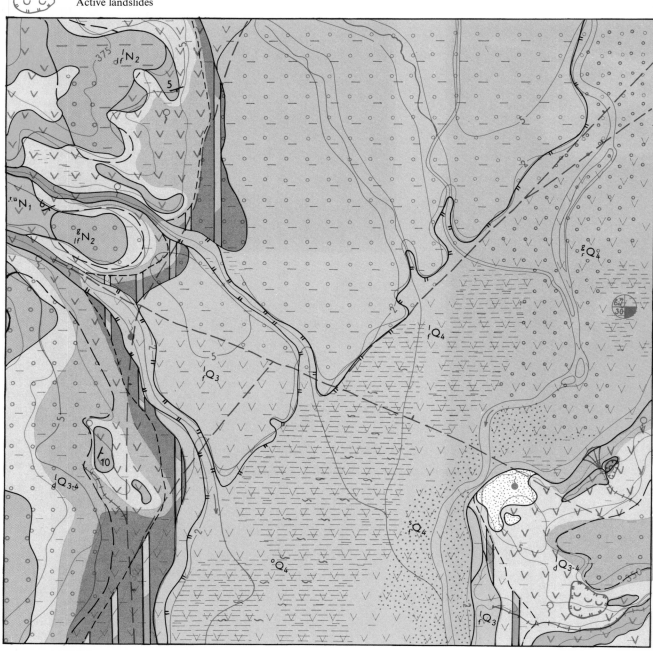

Comments

This map was set up for the purposes of land-use planning and designing common structures and engineering works. It comprises two parallel sheets, the map of engineering geological conditions (5.2.2.2) and the map of engineering geological zoning (5.2.2.3).

In the map of engineering geological conditions rocks and soils are divided into lithological complexes, which are distinguished by similar physical and engineering characteristics. Rocks and soils occurring at the surface are presented in a colour pertinent to a respective lithogenetical complex and in an orange pattern which indicates lithological character.

Three-dimensional representation using the stripe-method is applied to show the character of two superimposed lithological units of Quaternary deposits. The thickness of these deposits < 2 m, 2-5 m and > 10 m is shown in three shades of the particular colour. The lithological character of the pre-Quaternary bedrock units, where overlain by superficial deposits, is shown in grey patterns. Thick patterns indicate the bedrock surface to 5 m in depth, lighter patterns are used for 5-10 m depth.

Hydrogeological conditions are represented in blue colour by hydroisobaths of a maximum seasonal groundwater level, in intervals of 2, 5 and 10 metres. Specific symbols indicate the corrosiveness of the groundwater.

Distribution, kind and intensity of geodynamic processes, landslides, gully erosion, forming of alluvial cones, are shown in red.

I. GENETIC-LITHOLOGICAL CLASSIFICATION

Age	Genetic group	Symbol	Labelling of rocks		Lithological complex
			On the surface thickness (m) <2 2–5 5–10	Under surface complexes	
Quaternary	Fluvial	$_f^lQ_{3-4}$			Sandy-loessic to loessic-clayey alluvial loams
		$_f^sQ_{3-4}$			Alluvial sands predominantly medium-grained with loamy intercalations
		$_f^gQ_{3-4}$			Fine- to medium-grained sandy gravels, loamy in older terraces
		$_oQ_{3-4}$			Sapropelic muds in abandoned river channels, sandy-loessic to clayey
	Slopewash (Deluvial)	$_d^lQ_{3-4}$			Sandy-loessic to clayey slope loams
	Chemical	$^{tr}N_2-Q_4$			Travertines, mounds and slope sheets
Pliocene	Lacustrine	$_{lr}^lN_2$			Lacustrine and alluvial clayey-loessic, sandy-loessic, rarely clayey loams
	Lacustrine fluvial	$_{lr}^gN_2$			Lacustrine and alluvial sandy to loamy gravels with sand beds
Miocene	Volcanic lacustrine	$^{tu}N_1$			Stratified andesitic tuffs and tuffites

(continued overleaf)

5.2.2.2 *continued*

II. ENGINEERING CLASSIFICATIONS [1]						III. ENGINEERING-GEOLOGICAL CHARACTERISTIC	
Symbol	Building foundation conditions	Road construction		Soil classification	Ease of excavation	For rock material of the principal lithological types (on undisturbed sam...	
		Foundation	Material for embankments			Lithological types	
$_f^l Q_{3-4}$	D_2	MV	MV	ML	1–3	Sandy-loessic loams (in the Hron flood-plain)	
	D_3			CL	1–3	Clayey-loessic loams (in older terraces)	
$_f^s Q_{3-4}$	C_2	V	VV	SM	2	Loamy sands	
$_f^g Q_{3-4}$	B_1	VV	VV	GW	4	Sandy gravels (Hron valley terraces)	
				GM			
$_o Q_{3-4}$	E_3	N	N	OH	2–3	Sapropelic muddy silts of abandoned river arms in t... floodplain [5]	
$_d^l Q_{3-4}$	D_2	MV	MV	CL	2–3	Sandy-loessic loam	
	D_3						
$^{tr} N_2 - Q_4$	A_2	N	MV	—	5	More compact travertine	
						Strongly porous travertine	
$_{lf}^l N_2$	D_2	MV	MV	CL	2–3	Loessic-clayey soil	
	D_3			ML		Sandy-loessic soil	
$_{lf}^g N_2$	B_1	VV	V	GV	3	Weathered sandy gravels	
				GM		Loamy gravels	
$^{tu} N_1$	A_2	V	MV	—	5–6	Andesitic psammitic tuffs Psammitic tuffites Psephitic tuffites	

cific vity g.cm^{-3}]	Dry bulk density ρd [g.cm^{-3}]	Porosity %	Consistency Ic: Rebound hardness (R. Schmidt)	Moisture content %	w_p %	w_l %	I_p % Indentation hardness [kgf. mm^{-2}]
9[2]	1.51 / 1.42–1.62[3]	43 / 39–47	0.6 / 0.1–0.9	30 / 25–32	24 / 21–30	37 / 32–44	14 / 10–22
1	1.57 / 1.30–1.78	40 / 36–51	0.8 / 0.4–1.3	29 / 15–35	29 / 18–35	41 / 30–54	16 / 7–24
5	1.30–1.70[4]	36–47		9–10	—		
5–2.68	1.80–1.90						
0	1.37	48	0.13	33	21	35	14
7	1.62 / 1.13–1.77	40 / 32–58	0.7 / 0.5–1.3	26 / 18–43	23 / 17–40	39 / 23–52	14 / 6–20
56	1.93	8					150
58	1.60	16					—
53	1.76 / 1.15–1.60	45 / 40–55	0.8 / 0.3–1:0	29 / 23–46	25 / 19–40	40 / 29–60	16 / 12–23
	1.67 / 1.30–1.70	37 / 30–50	0.9 / 0.7–1.1	20 / 14–39	20 / 15–24	36 / 22–48	15 / 6–24
68	1.68	36	16		—		48
71	1.69	37	26		—		70
61	1.85	29	—		—		—

(continued overleaf)

5.2.2.2 *continued*

III. ENGINEERING-GEOLOGICAL CHARACTERISTIC *(cont.)*

For rock material of the principal lithological types (on undisturbed samples) *(cont.)*

Symbol	Shearing strength φ in degrees	Cohesion c in kgf.cm^{-2}	Uniaxial compressive strength σf [kgf.cm^{-2}]	Modulus of compressibility [kgf.cm^{-2}]			Others
				Modulus of elasticity E $(10^{-4}$ kp.cm$^{-2})$	Modulus of deformation E_o	Modulus of dynamic elasticity E_{dyn}	
$^l_f Q_{3-4}$	25	0.5	—	—	85	—	
	11	0.8	—	—	—	—	
$^s_f Q_{3-4}$	27	0.1					
$^g_f Q_{3-4}$	30	36					
$_o Q_{3-4}$							
$^l_d Q_{3-4}$	21	0.8	—	—	102–140	—	
$^{tr}N_2 - Q_4$			302			84	$k_z = 0.87^6$
			118			—	$k_z = 0.60$
$^l_{lf} N_2$	9	1.0			80		
	14	0.7			147–105		
$^g_{lf} N_2$							
$^{tu} N_1$			115–171	4	2.8	5	$k_z = 0.61–0.71$
			225	10.9	4.3	15	$k_z = 0.78$
			115	—	—	—	$k_z = 0.38$

1. Classification according to ČSN (Czechoslovak Building Standard): A_2 = soft rocks, B_1 = gravels with pebbles in contact, C_2 = medium to fine sands, D_2 = clays of medium plasticity, D_3 = clays of high plasticity; E_3 = organic silky clays, VV = very suitable, V = suitable, MV = barely suitable, N = unsuitable.
2. Mean statistical values.
3. Medium and minimum-maximum values.
4. Minimum-maximum values.
5. Analogous values from neighbouring area.
6. k_z coefficient of softening (ratio of saturated to dry strength).

r rock complexes (*in situ*)

ickness 0–2 m. Horizontal thin to laminar bedding. Vertical anisotropy. Strong slaking. Soft to solid consistency. Intercalations of organic soils. ght permeability.

ickness 0–1.5 m. Medium sands, intercalations of silts, cross-bedded. Considerable loam admixture decreases the permeability and increases herence. Facially very variable. Strongly weathered in older terraces.

ndy medium gravels with sand intercalations 3–9 m thick. Irregular bedding, vertical anisotropy. Moderate permeability 5.10^{-4} m/s. Water-bearing, cally corrosive. Boulders at the base. In older terraces strongly weathered, loamy, permeability 3.10^{-6}. In lateral valleys coarse, bouldery, terogeneous.

eposits occupy numerous irregular abandoned river arms and depressions at the edges of alluvial cones. Sandy and loessic sapropelic muds, ry soft to soft consistency, water saturated. Very low permeability, slakes readily. No bearing capacity, very strongly compressible.

regularly to cryptobedded, predominantly 1.5–3 m thick, at the foot of slopes more than 10 m thick, sandy-loessic brown loams, with fragments volcanic rocks. On Pliocene deposits more clayey, rust-brown, with gravels. Stiff to very stiff, weather to prismatic fragments. In places slake adily. Very slightly permeable. Used for brick making.

egular bodies, accumulations and sheets adapted to the relief. Variable thickness. Semi-solid rocks, strongly fissured, macroporous and permeable. luble, karstified. Hardly suitable for building or decorative stone.

cially very homogeneous cryptobedded clayey to sandy-loessic soils, up to 10–15 m thick. Brown to rusty-brown, spotted. Stiff to very stiff nsistency. Diagenetically consolidated. Prismatic disintegration. Practically impermeable.

ndy-loamy medium gravels 5–50 m thick. Lenticular intercalations of sands. Strongly weathered, in places weakly cemented. Medium permeability.

arse-bedded psephitic-psammitic tuffs deposited in water environment. They alternate with psephitic to pelitic tuffites. Varied colours, facial riability and physical heterogeneity. Weak semi-solid rocks, weather readily, disintegrate in water. Little jointing and low permeability.

5.2.2.3 *Multipurpose comprehensive, medium-scale map*

Map of engineering geological zoning, Zvolen Basin 1:25,000

Legend

—5—	Hydroisobaths of the max. groundwater table
○ ●	Springs of fresh and mineral water
☰	Waterlogged territory
pH ⊕ CO_2 / H ⊕ SO_4	Corrosiveness of groundwater: pH degree, hardness of water (in German degrees); the full sector indicates the aggressivity of CO_2 and/or SO_4
⬭	Spa-protection zones
⬭	District affected by landsliding

⌁→→	Districts affected by intensive gully erosion
——	Boundaries of zones
– – –	Boundaries of subzones
Fu, Dm	Symbols for zones
h2g2	Symbols for subzones
a / b	Pre-Quaternary soils with the surface in the depth (a) <5 m, (b) 5–10 m
a / b	Pre-Quaternary semisolid rocks with the surface in the depth, (a) <5 m, (b) 5–10 m

Comments

While the previous map presents information on the distribution and character of individual principal components of the engineering geological environment (rocks and soils, groundwater, geodynamical process), a comprehensive evaluation of these interrelated components at different places within the map area is presented in the map of engineering geological zoning.

The method of typological zoning is applied to delimit different types of map zones and subzones. The discrimination of engineering geological zones is based on the uniformity in general lithological character and structural arrangement of lithological complexes in the uppermost parts of the ground. Engineering geological subzones are delimited with-

in the zones on the basis of homogeneity in spatial (superposition) and proportional (thickness) arrangement of individual types of soils and rocks in schematized type cross-sections of the foundation soils.

Individual zones are indicated by symbols, expressing the genetical-lithological character of rocks (for example, Fu = zone of river valley deposits, Ft = zone of deposits in river terraces, Dm = zone of slope-wash deposits on magmatic or metamorphic rocks, Ep = zone of aeolian sands, etc.; there are forty types of zones adopted in the zoning classification in Č.S.S.R.). Should it be deemed necessary to take into consideration two superimposed complexes of smaller thickness, it is possible to indicate such zones also by combined symbols (for example, EsFt = zone of river terrace deposits overlain by aeolian loess).

ZONES				SUBZONES		
Type	Geological-geomorphological characteristic	Hydrogeological conditions	Geodynamic conditions	Type	Pattern (on the map)	Lithological characteristic Vertical lithological composition
Ng	Pliocene gravels alternating with sand horizons forming a flat rolling landscape with moderate slopes. Slopewash loams and sands with some gravel <1 m thick	Groundwater usually more than 5 m deep. Small springs from gravelly horizons	Extensive gully erosion. Slides of small extent	G3		Gravels. sandy gravels, less loamy gravels, with sand and clayey-sandy horizons
Dm	Slopewash loams 5–8 m thick on andesitic agglomerates and tuffs. Moderate slopes, locally intensively dissected by erosion	Very slightly water-bearing slopewash on relatively permeable bedrock	Extensive sheet and gully erosion. Shallow landslides and earth flows	$h2B^1$		Slopewash loams, clayey, less sandy, stiff to very stiff, 2–5 m thick, on agglomeratic tuffs and agglomerates, interbedded with lapilli tuffs
Fu	Floodplain alluvium formed of gravel beds 5–9 m thick covered by sand, or loam 0.5–2 m thick	Territory regularly affected by large floods. Water table usually <2 m deep. Water locally corrosive (high SO_4 content)	In places the river banks are undercut	$k1g3B^2$		Loam and sand 0.5–2 m thick, underlain with sandy gravels 6–9 m thick. In the depth of 5–9 m substratum as in zone Dm (tuffs)

Note: In this table the characteristics for only three selected types of subzones of three typical zones are presented as an example.

45

Subzones are quasihomogeneous models of qualitative and quantitative vertical structure of zonation grounds and their delimitation is made according to the following criteria (abbreviated):

Quaternary deposits

g = Gravelly soils
p = Sandy soils
n = Alternation of gravelly and sandy soils
h = Cohesive soils
k = Combination of cohesive and uncohesive soils
s = Loess soils
o = Organic soils
b = Bouldery soils

Pre-Quaternary basement

S = Solid (hard) rocks
B = Semisolid (weak) rocks
F = Alternation of hard and weak rocks (flyschoid)
Z = Highly weathered rocks
G = Gravelly soils
P = Sandy soils
N = Alternation of gravels and sands
I = Cohesive soils
K = Combination of cohesive and non-cohesive soils

Strata thickness

1 = <2 m
2 = 2–5 m
3 = >5 m

Depth of pre-Quaternary surface

1 = <5 m
2 = 5–10 m
3 = >10 m

Subzones are indicated by symbols, which are formed by grouping the corresponding signs for soil and rock (type, thickness, or depth of the pre-Quaternary basement) according to vertical sequence of delimited strata. For example the symbol $hlg2S^1$ expresses the model of the foundation soil, in which cohesive soils (thickness <2 m) are underlain by gravels (thickness 2–5 m) and in the depth to <5 m hard rocks of pre-Quaternary basement occur.

ENGINEERING-GEOLOGICAL AND GEOTECHNICAL CHARACTERISTICS OF THE SU

Estimated bearing capacity q_o kgf/cm^{-2} at a 2 m/6 m depth for foundation width 1 m. Depth of water below foundation:		Pile bearing capacity. Piles $\varnothing$1,000 cm^2, length 6 m	Foundation evaluation according to the standard for foundation	Suitability for road foundation according to standard for transport roads		Capillarity	Frost susceptibility	Evaluation of road construction conditions
				Road on natural subgrade	Embankment on natural subgrade			
>1 m	<1 m	(Mp)						
$\frac{7}{9}$	$\frac{4.7}{6}$	60	Foundation conditions simple. Foundation site suitable. Only conditionally suitable in loamy gravels. Gravels and sandy gravels, class 8 and 10. Loamy gravels, class 9 and 11	Very suitable	Very suitable	None (medium)	None (medium)	Territory suitable for roads in cuttings and embankments
$\frac{1.2}{6}$	$\frac{12}{6}$	60	Foundation conditions simple. Loams, class 20 and 21. Substratum class, 2 to 5. Foundation sites conditionally suitable	Barely suitable	Barely suitable	High	High	Because of soi character and extensive dissection, territory barely suitable for transport road
$\frac{6}{8}$	$\frac{4}{5}$	40	Foundation conditions simple. Foundation sites may be unsuitable because of high water or flooding	Stratum I (loamsand) little suitable. Stratum II (gravel) very suitable	Stratum I (loamsand) little suitable. Stratum II (gravel) very suitable	Medium to high	Medium to high	Because of hydrogeological conditions roads should be built on embankments

For the abbreviation of symbols (mainly in small sub-zones) on the map the signs of the pre-Quaternary basement are left out from the symbols. Instead the character and depth of the basement on the map is shown by grey patterns in different weight.

Apart from the division of the map territory into zoning units with approximately similar engineering geological conditions, the information for the user is completed by presenting also data on hydrogeological conditions. The parts of the territory affected by certain geodynamic processes are delimited as particular districts. Indicated also are various protected areas.

Maximum slope angle in temporary excavations. Depth $\frac{3m}{4m}$		Estimated maximum flow in l/min. through 1 m of the circumference of the excavation to lower the water table by:			Ease of excavation	Possible rock use	Recommendations for additional investigations
dry	Under water	2 m	4 m	6 m			
$\frac{1.4}{1.8}$	$\frac{1:2.2}{1:2.5}$	5	15	50	2–3	Gravel and loam extraction; gravels of poor quality	Because of the irregular occurrence of sand and clayey-sand intercalations a dense network of boreholes is required
$\frac{0.7}{0.9}$	$\frac{1:1.3}{1:1.2}$	—	<1	<1	2–3	—	Very variable petrographic composition and consistency of loams
$\frac{1.4}{1.8}$	$\frac{1:2.2}{1:2.5}$	40	180	400	$\frac{1–2}{2–3}$	Considerable quantities of a good quality fluvial gravel for concrete	Variable corrosive ground water properties

5.3   Examples of special purpose engineering geological maps

5.3.1   SPECIAL PURPOSE ANALYTICAL MAPS

5.3.1.1   *Special purpose, analytical, small-scale map*

Map of relative ease of excavation, Utah (United States), at a scale of approximately 1 : 200,000

Legend

1	Excavation very easy
2	Excavation easy
3	Excavation easy to difficult; variability due to inter-bedded resistant and soft rocks
4	Excavation difficult
5	Excavation very difficult

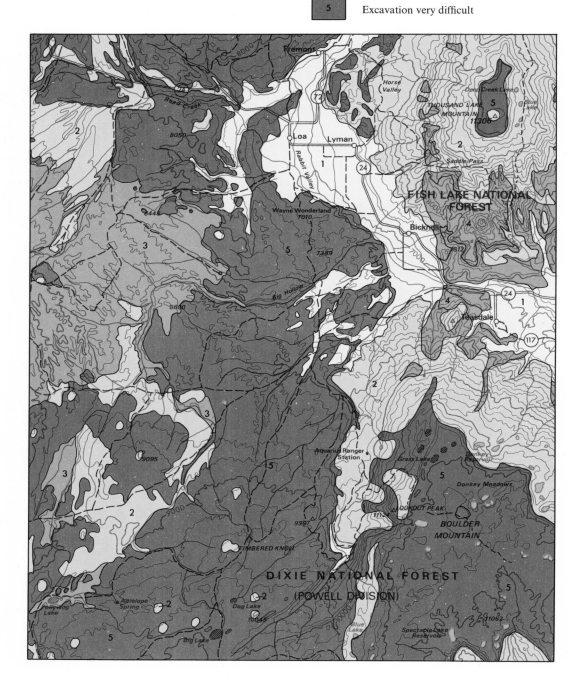

This map shows relative ease (or difficulty) with which rocks and surficial deposits can be excavated. Because of rapidly changing technology of excavation and considerable local variability of many rock units, it is not practical to specifically categorize rock units according to type of equipment needed for their excavation. However, it may be stated in general that rock units classed as *very easy* and *easy* can in most places be excavated by hand tools and by light machinery such as backhoes and small bulldozers; units included in *easy to difficult* require blasting and (or) heavy machinery such as rippers and large bulldozers for resistant rocks, and hand tools or light power equipment for soft rocks, and units classed as *difficult* and *very difficult* probably require blasting and heavy machinery.

The excavation units shown here are based on map units of the geologic map of the Salina quadrangle. Where bedrock is mantled with thin unmapped surficial deposits, ease of excavation shown is that of the bedrock, not that of the thin surficial mantle; where surficial deposits are mapped, ease of excavation shown is that of the surficial deposits.

Comment

The map selected here as an example of a special purpose, analytical, small-scale map is one of a series of maps prepared for the same area at a scale of 1:250,000. Map I–591 (sheet 1 of two) is a standard type of geological map with an extensive descriptive legend of surficial and solid formations. It serves to emphasize the close relationship between structure and stratigraphy and the derived 'ease-of-excavation' map illustrated here.

Other complementary maps in the series provide an example of the use of structural contours (map I–591, sheet 2 of two). Hydrological and hydrogeological maps include map I–591–D, showing normal annual and monthly precipitation in the Salina Quadrangle, Utah; map I–591–F, surface water, map I–591–G, springs; map I–591–K, general chemical quality of groundwater; map I–591–M, general availability of groundwater; and map I–591–N, drainage basins and historic cloudburst floods. Other specialized maps include map I–591–E showing the length of freeze-free season, and map I–591–H which shows the distribution of different types of bedrock and surficial deposits. The latter is a lithological map showing twenty-one lithological groupings with short descriptions, in the place of the eighty-three formations and other lithostratigraphical types depicted on the geological map. Basically map H is a simplified version of the main geological map, and illustrates the difference between lithological maps and lithostratigraphical maps.

This example has been redrawn from a part of P. L. Williams, *Map Showing Relative Ease of Excavation in the Salina Quadrangle, Utah*. Folio of the Salina Quadrangle, Utah Map I–591–J. United States Geological Survey, 1972.

5.3.1.2 *Special purpose, analytical, medium-scale map*

Map of foundation suitability of soils near Marseille (France)
reproduced at a scale of 1 : 35,500

Comment

The complete map, covering an area of approximately
1,000 km², indicates the suitability as foundation material
(bearing capacity) of the soil in general terms for depths of
0-3 m, 3-6 m and 6-9 m. The soils in the three layers have
been classified as good or very good (GVG) in green, good
on the whole (G) in yellow, bad and locally mediocre (B)
in red, and very bad and fill material (VB) in violet.

No documentation details are given, and it is not known
whether the mapping has been based on physical and
mechanical tests or on lithological description of the rocks.
On the complete map sheet the isopiestic lines for the watertable of La Crau are given; these are the only hydrogeological data given (Sanejouand, 1972, p. 32).

GOLFE DE FOS

PORT St LOUIS
DU RHÔNE

Recognition of four qualities of soil and three distinct layers makes it theoretically possible to show sixty-four different zones on the map. Although the system is very flexible, it is doubtful whether it is realistic to give regional estimates for layers of constant thickness, and whether the small scale of the map justifies recognition of three depth zones. In addition, the printed map is not particularly legible. However, the cartographic solution to the problem is not without interest as a method of possible general application in appropriate situations.

The map is part of the 1 : 50,000 map of the *Bassin de la Crau et de l'Etang de Berre, Carte d'Aptitude des Sols aux Fondations* prepared in 1969 by the Laboratoire Régional des Ponts et Chaussées de Marseille for the Organisation Régionale d'Etude d'Aire Métropolitaine 13.

Legend

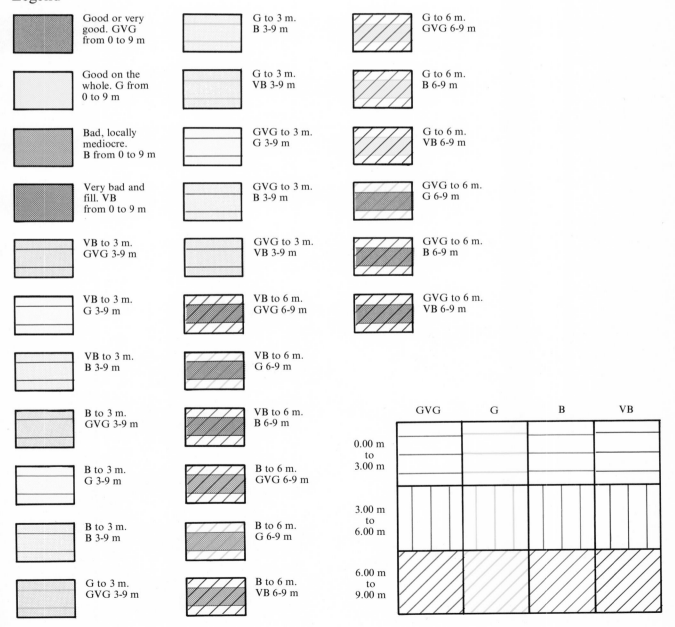

	Good or very good. GVG from 0 to 9 m
	Good on the whole. G from 0 to 9 m
	Bad, locally mediocre. B from 0 to 9 m
	Very bad and fill. VB from 0 to 9 m
	VB to 3 m. GVG 3-9 m
	VB to 3 m. G 3-9 m
	VB to 3 m. B 3-9 m
	B to 3 m. GVG 3-9 m
	B to 3 m. G 3-9 m
	B to 3 m. B 3-9 m
	G to 3 m. GVG 3-9 m
	G to 3 m. B 3-9 m
	G to 3 m. VB 3-9 m
	GVG to 3 m. G 3-9 m
	GVG to 3 m. B 3-9 m
	GVG to 3 m. VB 3-9 m
	VB to 6 m. GVG 6-9 m
	VB to 6 m. G 6-9 m
	VB to 6 m. B 6-9 m
	B to 6 m. GVG 6-9 m
	B to 6 m. G 6-9 m
	B to 6 m. VB 6-9 m
	G to 6 m. GVG 6-9 m
	G to 6 m. B 6-9 m
	G to 6 m. VB 6-9 m
	GVG to 6 m. G 6-9 m
	GVG to 6 m. B 6-9 m
	GVG to 6 m. VB 6-9 m

	GVG	G	B	VB
0.00 m to 3.00 m				
3.00 m to 6.00 m				
6.00 m to 9.00 m				

5.3.2 SPECIAL PURPOSE COMPREHENSIVE MAPS

5.3.2.1 *Special purpose, comprehensive, small-scale map*

Map of part of the Northeast Corridor, Washington, D.C., to Boston (United States), reproduced at approximately 1:265,000; produced for transport planning

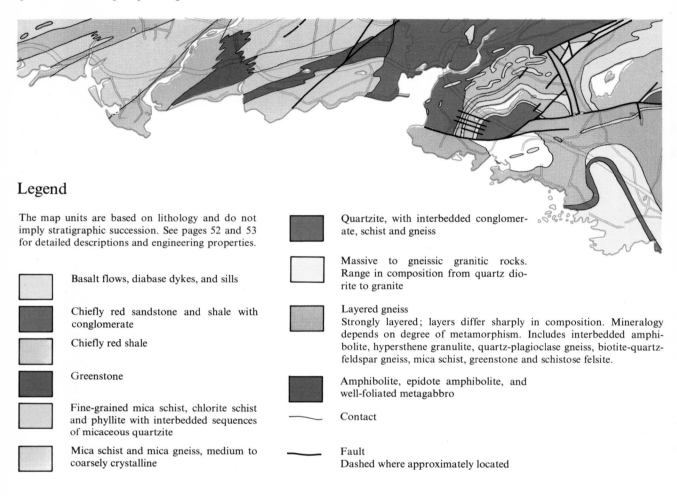

Legend

The map units are based on lithology and do not imply stratigraphic succession. See pages 52 and 53 for detailed descriptions and engineering properties.

Basalt flows, diabase dykes, and sills

Chiefly red sandstone and shale with conglomerate

Chiefly red shale

Greenstone

Fine-grained mica schist, chlorite schist and phyllite with interbedded sequences of micaceous quartzite

Mica schist and mica gneiss, medium to coarsely crystalline

Quartzite, with interbedded conglomerate, schist and gneiss

Massive to gneissic granitic rocks. Range in composition from quartz diorite to granite

Layered gneiss
Strongly layered; layers differ sharply in composition. Mineralogy depends on degree of metamorphism. Includes interbedded amphibolite, hypersthene granulite, quartz-plagioclase gneiss, biotite-quartz-feldspar gneiss, mica schist, greenstone and schistose felsite.

Amphibolite, epidote amphibolite, and well-foliated metagabbro

— — Contact

▬▬ Fault
Dashed where approximately located

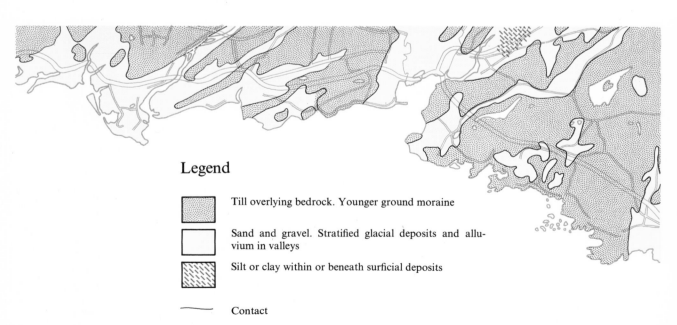

Legend

Till overlying bedrock. Younger ground moraine

Sand and gravel. Stratified glacial deposits and alluvium in valleys

Silt or clay within or beneath surficial deposits

— — Contact

Comment

The illustration is based on a small part of sheet 4 of five map sheets, prepared in 1967 by the United States Geological Survey at the request of the United States Department of Transportation of the Northeast Corridor, between Washington, D.C., and Boston, Massachusetts. Published map scale is 1:250,000.

The upper map segment shows the bedrock geology; the lower map is of the coastal plain and surficial geology. A three-part report (map I–514) consists of a map of the bedrock geology with geological cross-sections (map I–514A, sheets 1 to 5), a table of geological descriptions and engineering characteristics of each geological map unit (sheet 6), a map showing sources of data (sheet 7), and a list of additional references. Map I–514B contains information on the coastal plain and surficial deposits, and map I–514C contains data on earthquake epicentres, geothermal gradients, and major excavations and borings that serve as sources of engineering data (documentation map) within the Northeast Corridor.

Typical entries in the tabular descriptions of geological and engineering characteristics of a bedrock map unit and a surficial deposit map unit are reproduced on pages 52 and 53.

BEDROCK MAP UNIT

Geological description

Map Unit. Amphibolite, epidote amphibolite, and metamorphosed gabbro.

Equivalent geological unit.[1] Baltimore gneiss (part); Brimfield formation (part); Glastonbury gneiss (part); Maltby lakes volcanics (part); Marlboro formation (part); Middletown formation (part); Putnam gneiss (part); Tatnic hill formation (part); unnamed units.

Lithology. Massive to banded, tough, strong amphibolite, amphibolite schist, amphibolite gneiss, hornblende gneiss; in part metagabbroic. Forms extensive thick layers, lenses and pods. Commonly epidote-bearing; much epidote-rich amphibolite gneiss, and pods of epidote; includes extensive thick light-red to pink garnet-rich, layers. In places schistose toward margins; locally intruded by quartz diorite dykes.

Structure. Structure variable; rock ranges from massive to gneissic. Large massive bodies rarely show folds, whereas smaller or layered bodies commonly are complexly folded. Joints commonly persistent, but vary greatly in spacing.

Weathering. Weathers to sticky red clay south-west of glaciated area; almost unweathered in glaciated area.

Topography. Hills to low mountains; rugged in places, especially near rivers.

Physical properties [2]

Dry unit weight (kg per cubic metre). 3,000–3,200.
Compressive strength.[3] High to very high.

Engineering characteristics

Evaluation of rock for construction.[4]

Young's modulus of elasticity.[5] High to very high.
Relative drillability.[6] 2.
Excavation characteristics. Overbreak reported moderate to excessive depending on orientation of excavation to major joint systems. Rock loads mostly slight. Swelling ground probable in wet shear zones. Contains fibrous minerals which may slow excavation by boring machine.

Hydrologic conditions [7]

Significant hydrolic features. Water occurs in weathered zone at shallow depth, probably not much below 45 m. Maximum yields from fault zones.
Permeability. Secondary. Primary fracturing; chemical weathering.
Depth of wells. Range: 10–230 m. Median: 47.5 m.
Yield of wells. Range: 2.2–675 litres per minute. Median: 54 litres per minute.

1. The stratigraphic nomenclature used in this report is that of the authors of the various data sources and does not necessarily conform with usage of the United States Geological Survey.
2. Physical data available only for some rock units; where data are lacking, the physical properties are inferred from comparisons with those of rock elsewhere that possess similar composition, structure and geological histories.
3. Classification is for uniaxial compressive strength of intact rock. Strength is reduced by physical defects and chemical alteration in rock; it may differ with respect to bedding, foliation or direction of principal residual stress.

Strength class	Range of compressive strength $(kgf/cm2)$
Very high	$>2,200$
High	1,100–2,200
Medium	550–1,100
Low	280–550
Very low	<280

4. Reported construction characteristics are limited in number. Evaluations of rock units, for the most part, are inferred from generalized conditions of structure, alteration, hydrology and state-of-stress. Specific conditions can change within short distance. More refined engineering evaluations must be based on more detailed knowledge of geological conditions.

5. Inferred for intact rock. Modulus is reduced by physical defects and chemical alteration in rock; it differs with respect to bedding, foliation, or direction of principal residual stress.

Modulus class	Range of static modulus of elasticity $(kgf/cm2)$
Very high	$>8.4 \times 10^5$
High	5.6×10^5–8.4×10^5
Medium	2.8×10^5–5.6×10^5
Low	7×10^4–2.8×10^5
Very low	$<7 \times 10^4$

6. Number 1 indicates rock most difficult to drill. Numbers increase with ease of drilling.
7. The well-yield data used here are based on public-supply and industrial wells in which the maximum potential of the aquifer was being developed.

SURFICIAL DEPOSIT MAP UNIT

Geological description

Symbol and designation. Qt. Younger ground moraine.

Lithology. Till (GM, GC, SM).[1] Chiefly an unsorted mixture of clay, silt, sand, gravel and boulders. Varies from a cohesive, moderately clayey material with embedded pebbles, cobbles and boulders (boulder clay) to a very-well-graded sand with gravel, cobbles, boulders and minor silt derived mostly from nearby bedrock. Stratification generally lacking, poor to crude where present. Generally very compact, firm and friable; upper few feet may be somewhat loose. Colour varies with weathering (oxidation) and with colour of dominant source bedrock. Generally grey when unweathered, buff to brown when oxidized; reddish-brown in areas of reddish-brown bedrock. Locally contains or underlies thin lenticular sand and gravel. Overlain by many small, thin swamps containing muck and peat, and stream courses with post-glacial alluvium that are not shown on map. Locally patchy; numerous bedrock exposures within unit not shown.

Thickness. Very irregular, generally less than 7 m. Unit normally thin to absent on hill crests and along major escarpments, thickness downslope to 15 m or more on lower slopes. In drumlins may be 45 m thick or more.

Alteration. Upper 1–18 m oxidized; most stones unweathered or nearly so.

Topography. A nearly ubiquitous mantle overlying the bedrock surface. Unusually thick till underlies streamlined hills, called drumlins, which may be as much as 45 m high; some drumlins have a core of bedrock.

Technical characteristics

Evaluation

Foundation conditions.[2] Bearing capacity generally good because of high density and poor sorting. Expansion negligible.

Excavation characteristics. Generally easy to moderately difficult to excavate with power equipment. Highly compacted till ('hardpan'), and very strong and bouldery till can be troublesome to excavate without special equipment.

Slope stability.[3] Cuts higher than 12 m generally require individual stability analysis. For lower slopes, 1.5 on 1 to 2 on 1 generally considered safe. Vertical slopes up to 4.5 m common, particularly in more cohesive clayey till.

General hydrological conditions[4,5]

Significant hydrological features.[4] Small domestic supplies obtained from dug wells. Unit functions as a confining layer over some bedrock aquifers.

Yield of wells. Range: 4.5–68 litres per minute; median: 27 litres per minute.

Coefficient of permeability. Range: 0.8–180 litres per day per square metre; median: —.

Specific capacity[6] (litres per minute per metre). Range: —; median: —.

Water-table conditions

Specific yield.[7] 5–17 per cent.

Free water content.[8] 60–210 litres per cubic metre.

Excavation permeability.[9] 5.5–190 litres per day per square metre.

---

1. Unified Soil Classification System adopted by Corps of Engineers, United States Army, *The Unified Soil Classification System*, Waterways Experimental Station, Vicksburg, Miss., 1953, Vol. 1, 30 p., 9 pl.; Vol. 2, 11 p., 1 pl. (Tech. Memo. 3–357.) Based on grain size, gradation, plasticity and compressibility of soil. Symbols assigned are approximate, based upon limited test data.

2. Bearing capacity (numerical values (tons per square metre) applied to qualifying adjectives): very poor = less than 11; poor = 11–43; fair = 43–86; good = 86–350; excellent = greater than 350. Compressibility = volume decrease in a soil mass in response to an external load. Expansion = volume increase that is a function of load, time, density, water content and type of clay minerals.

3. Cut slopes (numerical values, in degrees, applied to qualifying adjectives): vertical = 90; near vertical = 80–89; steep = 45–80; moderate = 30–45; gentle = 0–30.

4. The well-yield data used in the preparation of this text are based on public-supply and industrial wells in which the maximum potential of the aquifer was being developed (litres per minute).

5. Data given under general conditions should be used for calculations pertaining to artesian aquifers in tunnelling or deep excavations where it may not be possible to dissipate the hydrostatic head.

6. Specific capacity is the discharge expressed as a rate of yield per unit of drawdown. Data used are selected to represent conditions of optimum well development and therefore reflect aquifer characteristics.

7. The ratio of the volume of water which a saturated rock or soil will yield by gravity to its own volume, stated as a percentage. Values used in this text are laboratory determinations or estimates based on field experience.

8. Litres per cubic metre yielded by gravity drainage.

9. Rate of flow of water in litres per day through a cross section of 1 square metre under a unit hydraulic gradient at prevailing water temperatures.

5.3.2.2   *Special purpose, comprehensive, medium-scale map*

Maps of Herceg Novy (Yugoslavia) at an approximate scale
of 1:2,300 showing lithology (upper map) and seismic microzoning (lower map) for urban planning

Legend

Geodynamic phenomena:

⎯⎯⎯⎯   Scarp of active landslide

⎯⎯⎯⎯   Upper limit of beach cliff and limit of wave
attack

Structural and tectonic phenomena:

⎯⎯⎯⎯   Boundary of engineering-geological complex

⎯⎯⎯⎯   Bedding, dip and strike

⎯⎯⎯⎯   Joint, dip and strike

═══   Fault, dashed where supposed

⎯⎯⎯   Minor fault, dashed where supposed

⎯⎯⎯   Reserve faults, major and minor, dashed
where supposed

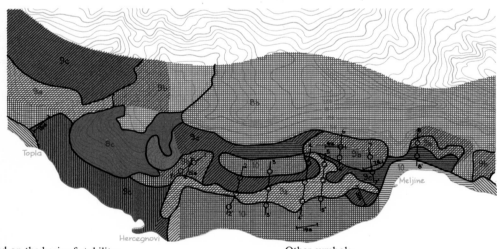

Legend

Terrain classified on the basis of stability:

▦   Unstable

▥   Unstable, partly stabilized

▨   Potentially unstable

▩   Conditionally stable (in conditions of predominantly horizontal relief)

▤   Stable

Other symbols:

o^{B4} Exploration borehole

▮   Sample for geomechanical analysis

⎯o⎯o⎯   Trace of electrical resistivity traverse

⊢⎯⊣   Trace of seismic refraction profile

Lithological composition and some important properties of the engineering geological complexes (top map).

KR Limestone and chert. Limestone with beds and lenses of chert, silicified marlstone and claystone
Limestone is thinly to thickly bedded, tectonically more fractured, karstification rather poor, porosity from fractured to cavernous, permeability irregular

KD Carbonate rocks—undivided. Limestone with subordinate dolomite, dolomitic limestone, sandy and marly limestone
Mechanically rather resistant, considerable karstification, cavernous to fractured, mainly readily permeable; rockfalls and talus on steep slopes

FK Flysch and flyschoid rocks. Marly limestone, conglomerate, sandstone, marlstone, claystone and transitional types, which occur rather irregularly
Alterations of platy, thin layered to laminated, rarely bedded to thickly bedded. Folded and fractured. Affected by surface weathering and erosion; fractures impart irregular porosity, permeability irregular

dl Talus. Carbonate stony debris, sand and clay with fragments and blocks of limestone
Talus has variable physical properties, prone to erosion and denudation processes, landsliding, etc.; unevenly consolidated, poorly graded and compacted, irregular porosity and permeability

Pr Torrential stream deposits. Gravels, sands and clays, mainly carbonate materials
Materials of variable grain size, poorly bedded and consolidated, irregularly permeable. Variable in thickness; contains organic matter

mn Littoral deposits. Boulders, gravels and sands, locally with clay; near the coast clay predominates
The deposits are very variable in grain size, poorly compacted, mainly saturated with water

as Alluvial beach deposits. Silts and clays, rarely sandy and gravelly; some detritus of weathered flysch
Deposits of variable grain size, poorly bedded, mainly compressible, saturated sediments with variable physical properties

nk Colluvial deposits. Very heterogeneous, mainly coarse detritus with blocks
Poorly graded, seasonally saturated with groundwater, irregularly permeable and liable to continual sliding

Map of seismic microzoning (bottom map):

8h Part of complexes of carbonate and limestones with cherts, tectonically moderately disturbed or in a zone of high slopes (K = 0.05)

8c More fractured and tectonically disturbed parts of complexes of carbonate and limestones with cherts, where large blocks are liable to fall (K = 0.06)

9a Complex of flysch and flyschoid rocks, small isolated masses of carbonate rocks on flysh, as well as highly consolidated semi-cohesive and uncohesive materials in the zone of surface weathering (K = 0.08)

9b Well consolidated parts of slope deposits, well graded and permeable, as well as proluvial and alluvial deposits where water table is deeper than 2 m (K = 0.10)

9c Poorly graded and predominantly impermeable parts of talus on slopes with inclinations greater than 12°, as well as parts of torrent and stream deposits of very heterogeneous composition, with water table less than 2 m deep (K = 0.12)

10 Highly compressible and saturated littoral deposits and colluvial deposits of variable permeability (K = 0.12)

Note: K = coefficient of seismicity

Comment

The single-map sheet, measuring 970 × 675 mm, is printed in colour. It includes two maps of the area at a scale of 1:25,000. The first, part of which is presented here in the upper illustration, is a lithological map with eleven horizontal cross-sections. The second, presented in part in the lower illustration, is a map of seismic microzoning on which the map units are defined in terms of coefficient of seismicity. As an overprint to this map terrain is classified according to stability.

Also provided on the map sheet is a small-scale, 1:100,000, map of seismic classification according to maximum intensity with epicentres for the period 1853–1970. Isolines are also given delimiting areas of different seismic intensity for the period 1667-1970.

A large-scale cross-section shows an interpretation of the structure of a selected area effected by multiple landsliding, and the bases for stability calculations.

The map sheet is an excellent example of the use of an extended descriptive legend, examples of which are reproduced here. Symbols for hydrological and hydrogeological features are also used on the published sheet.

Extract from the *Engineering Geological Map of the Urban Area of Herceg Novy, Jugoslavia*, compiled at a scale of 1:25,000 by D. Gojgic and M. Lazic in the Institute for Geological and Geophysical Research, Beograd, 1971.

5.3.2.3 *Special purpose, comprehensive, large-scale map*

Map of a mining area, originally produced at 1:10 000, reproduced here at 1:100,000 approximately

Comment

The map illustrated here is one of a pair which shows the engineering and geological conditions of surface strata; the other (Golodkovskaja and Demidyuk, 1970, Fig. 2) provides information about the engineering geology of the bedrock.

This map, of an area underlain by a mineral deposit in a permafrost region, has been used in the design and construction of mining and industrial works, urban, road and other structures required for the development of the deposit. It shows the age and genesis of the uppermost and underlying deposits, their lithology, thermal condition, frozen water (ice) content, range of average annual rock temperature, depth of seasonal and permanent frozen ground, depth of

frozen and partly frozen water, thickness of Quaternary deposits, and contemporary geological processes and phenomena. It also gives an evaluation of the area for construction purposes.

Despite the abundance of symbols the map is easily understood in this black-and-white version; legibility could be improved by the use of colour.

From the point of view of engineering-geological cartography, the map illustrates an interesting application of the 'stripe method' (Section 4.5) of indicating the nature of deposits underlying the uppermost beds combined with other information. Shading in the stripes (symbols 7 and 8) illustrates the distribution of marine Upper Quaternary and Permo-Triassic rocks; the direction and width of the stripes (symbols 9 to 12) indicates the average annual temperature range of the rocks; while the three degrees of spacing of the stripes (symbol 13; here illustrated as applied to symbol 11(a), but applicable to the whole range of symbols 9 to

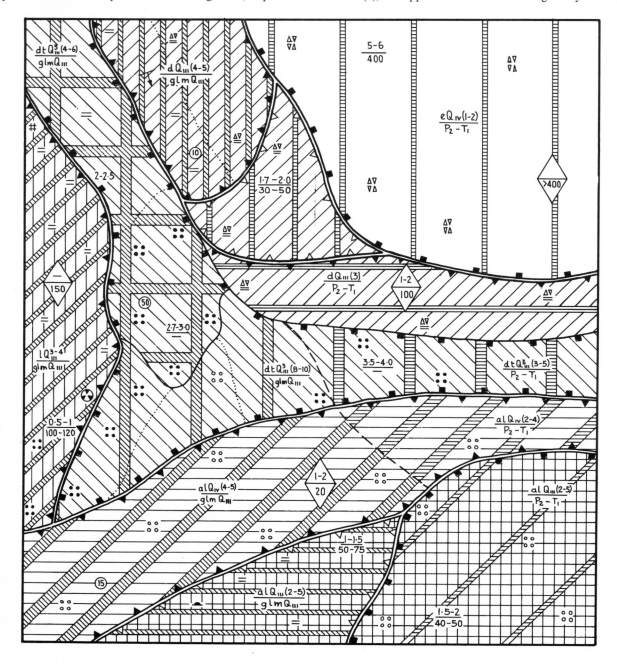

12 as shown on the map) indicate the frozen water (ice) content of the rocks.

A map of this type has reached the practical limit of complexity and would need to be accompanied by complementary hydrogeological and topographical maps (2.3.2.4) and auxiliary maps showing for example the composition and thickness of the surficial deposits (2.3.2.3).

The original version of this map was published as Figure 1: 'Montage of Engineering and Geological Map of Quaternary Deposits with Elements of Territorial Engineering and Geological Evaluation for Ground Construction', to accompany the paper by G. A. Golodkovskaja and L. M. Demidyuk, 'The Problems of the Engineering and Geological Mapping of Deposits of Mineral Resources in the Area of Eternal Frost' *Proceedings of the first International Congress of the International Association of Engineering Geology, Paris*, 1970, Vol. 2, p. 1049-68. The version published here has been redrawn and slightly modified.

Legend[1]

Genetic types and age of deposits:

1 — Eluvial, contemporary (eQ_{IV})

2 — Recent alluvium (alQ_{IV})

3 — Deltaic, Upper Quaternary (dtQ_{III})

4 — Alluvial, Upper Quaternary (alQ_{III})

5 — Diluvial, Upper Quaternary (dQ_{III})

6 — Lacustrine, Upper Quaternary (lQ_{III})

7 — Glacial, marine Upper Quaternary ($glmQ_{III}$)

8 — Volcanogenic Permian-Triassic (P_2-T_1)

Average annual range of temperature of rocks:

9 — —0° to —1°

10 — (a) —1° to —2° (b) +1° to +2°

11 — (a) —3° to —4° (b) +3° to +5°

12 — +5° to +7°

Frozen water (ice) content of rocks:

13 — (a) Low (b) Medium (c) High

Lithology:

14 — Rock debris with loam matrix

15 — Clay, loam with rock debris

16 — Inequigranular sand

17 — Pebbles and sand matrix

18 — Clay

Recent geological processes:

19 — Formation of fissures under conditions of frost

20 — Swelling

21 — Melting of ice intrusions

22 — Solifluction

Other symbols:

23 — Depth of seasonal melting in metres (numerator) Depth of permanently frozen ground in metres (denominator)

24 — Depth of frozen water (numerator)

25 — Thickness of Quaternary deposits

26 — Temperature boundaries

27 — Boundaries of underlying rocks

28 — Lithological boundaries

Boundaries of engineering and geological areas:

29 — Requiring no special engineering preparation

30 — Requiring engineering preparation

31 — Requiring very complicated engineering preparation

---

1. Note that the symbols in the legend have been reduced more than the equivalent map symbols.

5.3.2.4   *Special purpose, comprehensive, large-scale map*

Map scale 1:1,550 approximately

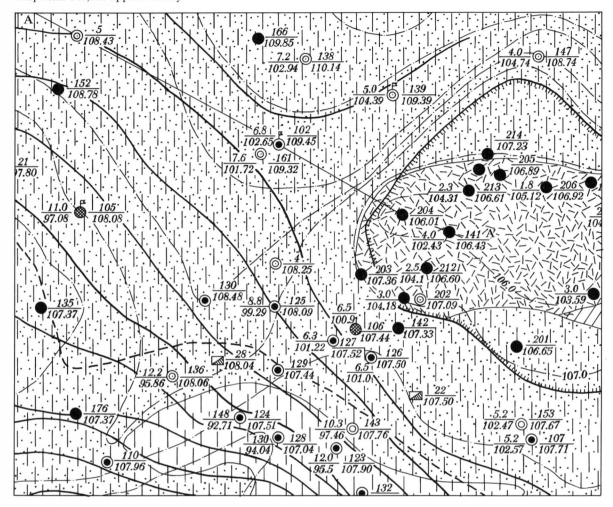

Cross-section A-A`. Engineering-geological cross-section (with indices of saturation, consistency, loess collapsibility, strength and deformability).

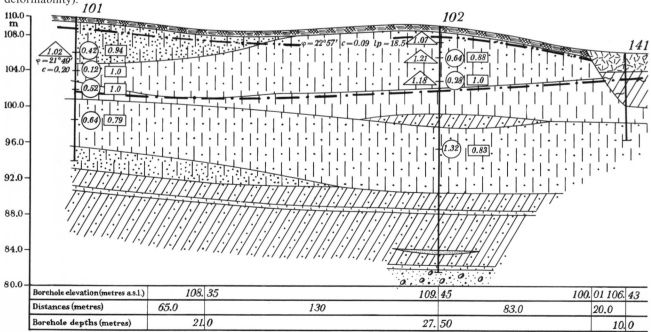

Legend

● Boreholes, 5–10 m in depth

◎ Boreholes, 10–20 m in depth

◉ Boreholes, 20–50 m in depth

⬤ Boreholes deeper than 50 m

Borehole with long-term observations of the groundwater table

Borehole for pumping tests

Old karst sinkhole

(0.64) Index of consistency (after Makeev)

1.0 Degree of saturation

△1.21 'K' coefficient of macroporosity (in loess materials)

φ Internal friction angle

c Cohesion, kg. cm^{-1}

lp Modulus of compressibility in mm.m^{-1} (loading at 2 kg.cm^{-2})

Surface isolines of the zone, where clayey materials have a plastic consistency

Groundwater table

Borehole profile, where samples were taken for laboratory tests

Southern boundary of Neogene

Hydroisolines on 25 September 19-, at 1 m intervals

A ——— A' Geological cross-section lines

$\dfrac{6.8}{102.65}$ ○ Depth of shallow groundwater (in metres) from the surface / Elevation of groundwater table (a.s.l.)

○ $\dfrac{161}{109.32}$ Borehole number / Elevation (a.s.l.)

⬚ $\dfrac{28}{108.04}$ Shaft number / Elevation (a.s.l.)

Surface contours (gradation of 0.5 m)

Top soil

IQ$_{III}$ {

Peat, greyish-brown

eQ$_3$ Peaty clayey silt, greenish-grey

Sandy silt, loess-like, yellowish-brown

Silt, loess-like, yellow-brown

alQ$_2$ Clayey silt, loess-like, yellowish-brown

alQ$_{II}$ {

Clayey silt, loess-like, grey

Sandy clay, greenish-grey

Basal conglomerate

Comment

The map is accompanied by an axonometric three-dimensional model, and tables of characteristics of map rock units and characteristics of delimited engineering-geological districts.

In the table of rock characteristics each distinctive engineering-geological type, shown in the map and cross-section, is described in detail as far as the lithological character and engineering classification properties are concerned (according to building standards), physical properties (grain size, specific gravity, bulk density, porosity, natural moisture content, degree of saturation, plasticity limits, consistency, collapsibility), and mechanical properties (angle of internal friction, cohesion, modulus of compressibility at 2 and 4 kg.cm^{-2} loading).

The table of characteristics of individual engineering-geological districts contains the following data: description of the lithological profile of foundation soils, hydrogeological conditions, geodynamic phenomena, collapsibility of materials, other special local characteristics, as well as engineering recommendations.

This is one of the first published engineering-geological maps, a pioneering effort which is very close in basic principles of mapping and presentation of data to the proposals in this guidebook. Made for the practical purpose of the engineering design of an industrial plant, the map was published as an example of mapping by I.V. Popov, R.S. Kats, A.K. Korikovskaya, and V.P. Lazareva in 1950 in their book *The Techniques of Compiling Engineering Geological Maps.*

Their map of engineering geological conditions, reproduced here in black-and-white, had, in the original, a territorial division into engineering-geological districts distinguished by colours.

5.3.2.5   *Special purpose, comprehensive, large-scale map*

Engineering geological map of Hannover at 1 : 6,350

Comment

A typical example of an urban engineering geological map built up from extensive borehole observations and other near surface observations. Particular attention is paid to areas of artificial fill and the natural surficial deposits of importance in foundation engineering. Solid geology and hydrogeological conditions, both natural and man-made, are also shown.

Parts of the engineering-geological (1:10,000) and groundwater (1: 20,000) sheets of the new engineering geological map of Hannover, 1970, here combined in a simplified single sheet at a scale of 1 : 6,350.

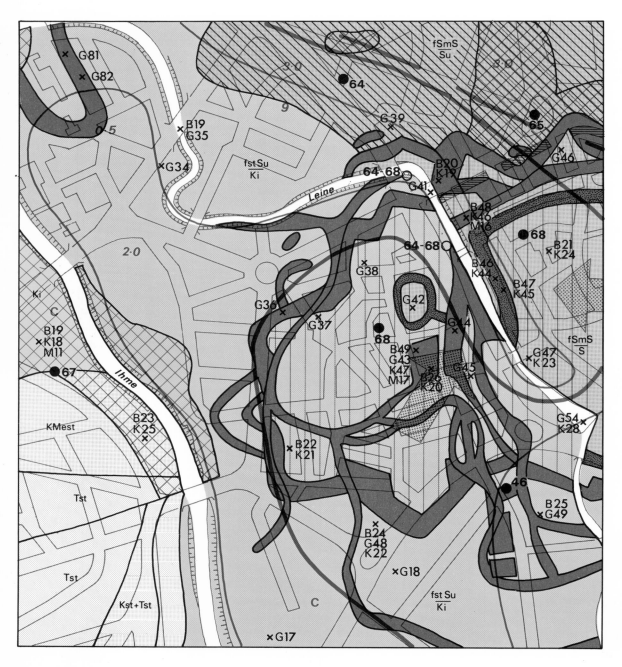

Legend

Infilling of the old town-moat (about 1550–80)

Abandoned meander of the Leine, ponds and fortification moats, silted up, in part infilled

Embankment, tipped and poorly compacted soil

General thin fill within the old town

Ki — Gravel (more than 2 m and up to 3 m thick), with few to many lenses of fine to coarse sand with fine to medium gravel (fluvio-glacial sand), dry to damp, round to angular, of varied lithology

$\frac{fSmS}{S}$ — Fine sand with a little medium sand (up to 2 m thick, round and equigranular: = dune sand) generally overlying alluvial sand, mainly dry

$\frac{fstSu}{Ki}$ — Alluvial loam, fine sand with clay (up to 5.5 m thick) uppermost metre generally clayey with deposits of peat (up to about 1 m) and sapropelic mud (up to 1.2 m thick) over gravel

$\frac{fSmS}{Su}$ — Fine sand with a little medium sand over alluvial loam

KMest — Marly limestone, compact, widely jointed, joints in part water-bearing; locally flaggy, Kca 1 + 2 (Campanian, Upper Chalk) on the geological map

Tst — Mudstone, compact (in even beds more than 100 m thick) with a covering of unconsolidated soil, the uppermost 1–2 m mainly stiff plastic clay, softened, generally widely jointed

×**K45**
M17
B47 — Sampling point for grain size (K), mineral (M) and soil mechanics (B) analysis

×**G43** — Site investigation

Highest known groundwater level contour, with height in metres above datum, based on many years of observation in good water-bearing beds, generally in sands and gravels in part interstratified with loam and clays

Approximate groundwater contours

Ancient principal main drainage channel, only used occasionally for drainage

3·0 — Thickness, or the sum of the thicknesses, of water-bearing strata

●**68** — Observation well (since 1941)

○**64-68** — Observation well (1943–64)

Direction of groundwater flow

C — Sample site for groundwater analysis

5.4 Interpretative geological maps

5.4.1 INTERPRETATIVE GEOLOGICAL MEDIUM-SCALE MAP

Legend

Qu	Qu	Undivided Quaternary deposits
	Pleistocene	
Qsu	Qsu	San Antonio formation. Upper member; clay, silt, sand, and gravel
Tl	Pliocene (?) T1	Leona Rhyolite
Jk	Upper Jurassic Jk	Knoxville formation. Shale, sandstone, and minor conglomerate

The map has been redrawn and slightly simplified from map GQ–769, *Areal and Engineering Geology of the Oakland East Quadrangle, California*, by Dorothy H. Radbruch 1969, United States Geological Survey, Washington, D.C. An explanatory pamphlet accompanies the map.

The map is reproduced here at a scale of 1:20,000.

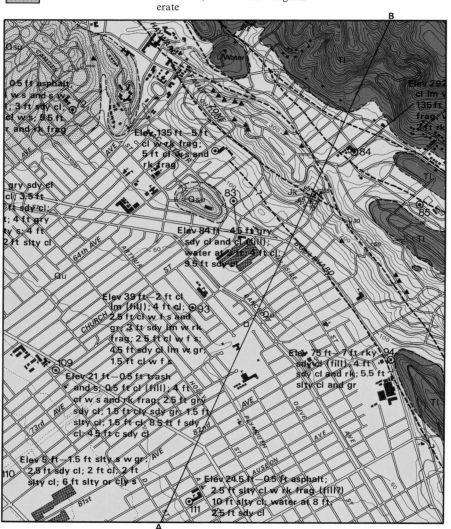

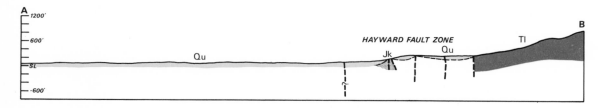

Legend

———————

Contact   Long-dashed where approximately located; colour boundary without contact line where hypothetical

————— ^{55} — ? — —

Fault, showing dip. Long-dashed where approximately located; queried where probable

_ ^{72} Strike and dip of beds

_ ^{65} Approximate attitude of beds determined from aerial photographs

Abbreviations

Elevations given below are to nearest foot. Thickness figures in logs are to the nearest 0.5 ft. Only the type of material encountered is given; descriptive details are omitted to conserve space. The following abbreviations are used in logs: ct, chert; cl, clay; cly, clayey; c, coarse; decomp, decomposed; f, fine; frag, fragment(s); gr, gravel; gry, gravelly; lge, large; lm, loam; mtl, material; mat, matter; med, medium; misc, miscellaneous; org, organic; pt, peat; peb, pebble(s); r, red; rk, rock(s); s, sand; ss, sandstone; sdy, sandy; sp, serpentine; sh, shale; shl, shell(s); si, silt; slty, silty; sm, small; tr, trace; ve, vegetable; w, with; weath, weathered.

▲
Landslide
The symbol marks the approximate location of the landslide. Size of the symbol bears no relation to areal extent of the slide. The landslides as located on the map constitute a record of past slope failures and are not by themselves proof of present or future slope instability.

Every landslide shown on the map was well defined and easily observable at the time of the study; however, not every landslide in the area is shown on the map. Therefore, although a landslide exists (unless it has since been removed by construction activity) at every point where one is shown on the map, it does not follow that there are no landslides elsewhere on the map.

83
⊙   Site of boring
Logs of borings are shown on the map where possible; where space does not permit showing the log on the map, it is given below. Abbreviations used in logs are explained below.
Number of item below refers to site number on map.
85. Elev. 383 ft. 9 ft sdy cl and c to f s (fill); 9 ft gr-s-cl mix; 6 ft cl; 11 ft sdy cl w tr f gr and si; 21 ft cl w s; 7 ft slty cl w f gr; 14 ft gr-s-cl mix.

□

Location of consolidation test

Generalized description of engineering properties of map units [1]

Map unit	General lithologic description	Topographic form
Undivided Quaternary deposits (Qu)	Composition and physical properties vary. Consist predominantly of Temescal formation. Probably include covered or unrecognized San Antonio formation and gravel, sand and clay (Qg), as well as recent alluvium and colluvium, and artificial fill. Symbols for Qtc, Qts, and Qtb shown in parentheses where these units can be positively identified (see Temescal formation)	Primarily in valleys and on gentle slopes between San Francisco Bay and the Berkeley Hills
San Antonio formation. Upper member (Qsu)	Clay, silt, sand, and gravel. Some pebbles soft; most firm. Most beds contain flakes and pebbles of white Claremont chert, some gravel almost entirely chert. Contains montmorillonite clay. Pale-yellowish-brown to greyish-orange. Consolidation varies, some layers loose, unconsolidated. Three consolidation tests on clay layers showed compression of 4 to 6 per cent. Maximum thickness unknown. May include some Temescal formation and lower member where exposures too poor to differentiate units	Primarily in rather steep dissected hilly areas between San Francisco Bay and steep front of Berkeley Hills
Leona rhyolite (Tl)	Rhyolite. Fresh rock light-grey to greenish- or light-bluish-grey, weathers to white or dark-yellowish-orange, may be iron-stained reddish-orange. Fresh rock contains abundant pyrite in many places. Contains a small amount of glass. Sheared and fractured. May include small amounts of Franciscan and Knoxville sandstone and shale too small to show on the map. Much of rhyolite apparently intrusive (Case, 1963); in places intruded overlying Knoxville shale, now baked and contorted at contact	Forms steep knobby dissected hills
Knoxville formation (Jk)	Shale, olive-grey, fissile; sandstone, fine- to medium-grained, olive-grey; also includes pebble conglomerate in dark shale or sandstone matrix, minor concretionary limestone, and lignite. Some shale massive, some interbedded with sandstone. Shale contains abundant *Buchia piochii*. Includes younger *Buchia*-bearing marine sedimentary rocks described by Case (1968). Thickness and stratigraphic relations unknown	Generally forms valleys, because soft shales of formation are easily eroded

Faulting: *The rocks of most of the above units have been compressed into northwest-trending folds and cut by numerous faults.*

The fractured rocks along any of the faults mentioned above may form passages for ground water, and cuts made across them may require draining; the soft sheared rocks are also subject to landsliding.

Severe earthquakes were caused by movement along faults within the Hayward fault zone in 1836 and 1868. Therefore, the entire length of the Hayward fault zone in this quadrangle can be assumed to be active.

Slow tectonic movement, or creep, is at present taking place at several locations along the Hayward fault zone, with resultant damage to manmade structures which cross the line of creep. Both the Claremont water tunnel and the drainage culvert under the University of California stadium have been damaged by this slow movement along a fault plane or band of shearing within the Hayward fault zone. It is not known whether creep is occurring along the fault zone elsewhere in this quadrangle, although discrepancies recently noted in rechecks of survey lines crossing the zone at 98th Avenue and at Lincoln Avenue may indicate right-lateral movement within the fault zone of approximately 0.1 to 0.15 ft in 10 years.

Structures which lie within or cross the Hayward fault zone may not only be damaged by sudden movement, offset, and rupture along a fault at the time of an earthquake originating in the fault zone, but may also be subject to constant strain and damage due to the opposite sides of faults within the zone continuously moving very slowly in opposite directions.

Weathering and soil development	Workability	Slope stability and foundation conditions	Dry density[2] moisture content and Unified Soil Classification[3]	Remarks (includes use and earthquake stability)
Soil may be as much as 3ft thick. *In places soil clayey, shrinks and swells; may cause damage to buildings*	Can be moved with hand or power tools	Depend on composition; generally good. *Slides have formed where colluvium apparently derived from gabbro*	Varies	Mapped with Temescal formation in Oakland West quadrangle (Radbruch, 1957)
Soil as much as 3 ft thick in places. *Soil swells and shrinks with seasonal moisture changes and may cause damage to buildings; may creep on slopes*	Can be moved with hand tools	*Large slides have formed in this unit.* Factors contributing to slide probably include presence of montmorillonite clay and alternating poorly consolidated sand and clay; steep slopes; and groundwater. Generally suitable foundation material for light structures where slopes are not steep	*1.68*; 18 per cent (77: *1.45-1.97*; 8-30 per cent) GM-CH	
Weathering as much as 30 ft deep; highly weathered rock consists of loose fragments in clay matrix. Soil generally lacking or less than 18 in thick; in ravines may be more than 12 ft thick	Can generally be moved with power equipment; in some places requires blasting	Slope stability and foundation conditions good. *Rare debris slides observed where rock excessively fractured and weathered*	*2.59* (s); 0.1 per cent; 99 (weathered); 20 per cent (3: *1.57-1.63*; 9-27 per cent)	Crushed Leona rhyolite is a major source of fill and base rock; pyrite formerly mined for sulfur; *runoff from rhyolite hills very acid and corrodes concrete sewer pipe. Some slopes so steep that development may be difficult*
Depth of weathering irregular; may be 20 ft or more in places. Some weathered rock firm, most soft, clayey. Soil commonly 1-3 ft thick	Can be moved with power equipment	Slope stability and foundation conditions generally fair; minor sloughing in cuts	*2.56* (s) (ss); 1.4 per cent; *1.86* (weathered sh); 15 per cent (3: *1.81-1.92*; 13-19 per cent)	*May squeeze in tunnels where sheared*

1. Text printed in italics indicates geologic conditions that may be critical to planning, design, and construction of engineering works.
2. Dry density (italic) expressed in tonnes per cu. m, based on one sample of fresh rock unless otherwise noted. Number of samples and range of dry density and moisture content given in parentheses (12: *1.69-1.74*; 17-20 per cent). (s) indicates sample collected at the surface. Moisture content (per cent) generally higher for subsurface samples of rocks than for those collected at the surface.
3. Unified Soil Classification (letter symbol) given where applicable (United States Army, Corps of Engineers, 1953, *The Unified Soil Classification System*; United States Army, Corps of Engineers, Tech. Memo. 3-357, Vol. 1-3).

5.4.2 INTERPRETATIVE GEOLOGICAL LARGE-SCALE MAP

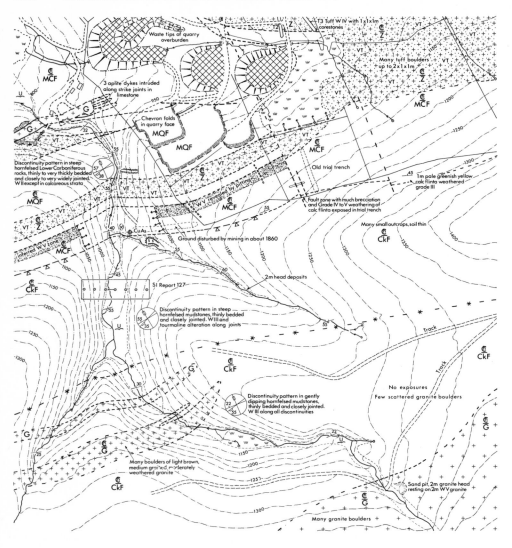

Rhyolitic tuff	Alluvium along stream	Boundary of superficial deposit, certain
Granite, fine-grained	Mine shaft, abandoned	Boundary, solid, certain
Granite, coarse-grained	Slope on spoil tip	Boundary, solid, approximate
Zone of weathering grade V	Solifluxion lobe	Axial trace of syncline
Spoil tip	Inclined strata, dip in degrees, normal succession	Fault, approximate position
Quarry	Inclined strata, dip in degrees, inverted succession	Trial trench
Stream, with direction of flow	Vertical strata	Site investigation
Spring	Joint, inclined, dip in degrees	Borehole
Seepage line	Joint, vertical	Borehole, inclined
	Discontinuity pattern, statistically determined	

Comment

The map, reproduced at a scale of 1:10,000, is an example of a geological map and legend supplemented with additional descriptive information in engineering geological terms. It is based on a part of a typical Institute of Geological Sciences map produced on a scale of 1:10,560; such geological maps are available for over 85 per cent of the land area of the United Kingdom and are useful at the preliminary planning stage of an engineering undertaking; mechanically enlarged, they could form the basis of pre-construction or site investigation maps. Supplementation was not undertaken officially.

This example has been slightly modified from 2.9.1. Part of a 1:10,560 map supplemented as proposed. Dearman *et al.* 'Working Party Report on the Preparation of Maps and Plans in Terms of Engineering Geology', *Q.Jl Engng Geol.,* vol. 5, 1972, p. 293–381.

Legend

SUPERFICIAL DEPOSITS (DRIFT). RECENT AND PLEISTOCENE

		Thickness (in m)
⌐ ~ ⌐	Alluvium. On the granite, alluvium is a brownish-yellow, loose, sub-angular, coarse gravelly sand with some peat and rounded boulders of moderately weathered granite up to 1 m, and pebbles of quartz. Downstream, alluvium is a silty gravelly sand with rounded granite boulders up to 1 m and sub-angular cobbles and boulders of the solid rocks. The deposits are moderately to highly permeable. Locally much disturbed by streaming for tin.	Up to 3
⌐ U ⌐	River terraces (undifferentiated). Dark yellowish-brown, loose but locally weakly to strongly cemented in horizontal layers by manganiferous or ferruginous material, sub-angular to rounded, sandy gravel with rounded to sub-angular cobbles and boulders of local rocks. Boulders occasionally up to 1 m. The deposits are highly permeable except where cemented. Locally much disturbed by streaming for tin.	Up to 12
⌐ ₵ ⌐	Head. Almost everywhere present and largely obscures the solid formations. Represents solifluxion debris and grades downslope into alluvium and terrace deposits.	2-3, locally >12

Within the outcrop of the granite, head comprises yellowish-brown, loose, layered, sandy gravel with some clay, and gravelly silty sand with cobbles and boulders of moderately weathered granite; grades down into moderately to highly weathered granite *in situ*. On the Upper Carboniferous outcrop next to the granite, head is typically reddish-brown, loose to compact, homogeneous, clayey gravelly sand with many sub-angular cobbles; on steep slopes fines may be absent and head is then loose, clean cobbles of the local rocks beneath 15–30 cm of humic soil.

On the Lower Carboniferous rocks, head is reddish-brown, loose to compact, homogeneous silty clayey sand with some cobbles and boulders of local rocks; it may be layered with an upper grey horizon separated by a black cemented layer typically 8 cm thick from reddish-brown head down to bedrock.

SOLID FORMATIONS. CARBONIFEROUS
Upper Carboniferous (Namurian)

		Thickness (in m)
CkF	Crackington formation. Dark to very dark grey, very fine grained, thinly bedded to thinly laminated, very closely jointed, slightly to moderately weathered, poorly cleaved shale, weak, impermeable except along open joints. Interbedded with very subordinate grey to dark greenish grey fine-grained, very thinly bedded, thinly laminated and cross-laminated, closely, jointed, slightly to moderately weathered siltstone, moderately strong and dark greenish grey medium grained, very thinly to medium bedded, with closely to widely spaced joints slightly to moderately weathered, sandstone, strong. The shale slakes on exposure and is suitable for brick making.	?
S	Sandstone. It has been possible to map groups of beds in which sandstone predominates. Beds are usually less than 30 cm thick and are separated by very thin beds of siltstone and shale. Sandstones are suitable for aggregate production. Within the contact metamorphic aureole of the granite, dark grey, very pale orange to dusky yellowish-brown, fine to medium grained, thinly bedded, closely jointed, slightly to moderately weathered, hornfelsed shale and sandstone, strong, impervious except along open joints. Locally with fine grained black tourmaline developed as selvedges up to 2.5 cm wide along discontinuities and with irregular quartz veins up to 5 cm wide.	

Lower Carboniferous (Dinantian)

⌐ ⌐	Meldon chert formation	75

5.5   Documentation maps

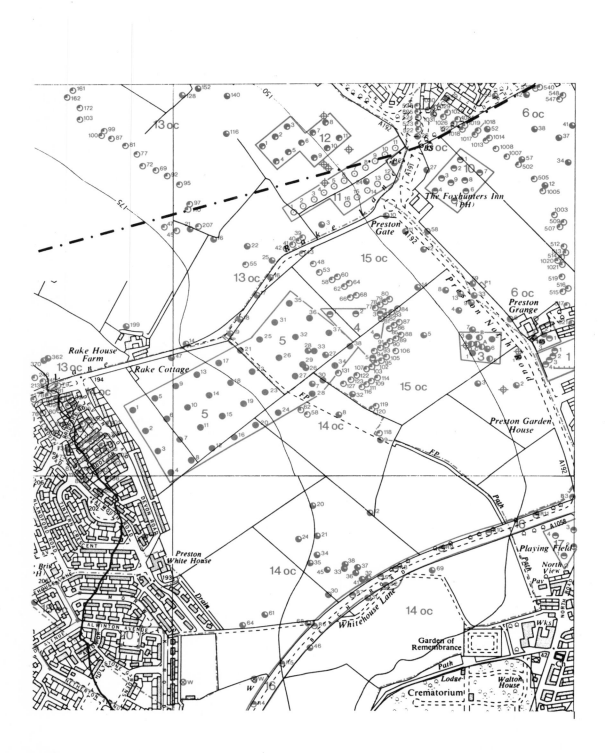

Legend

Opencast prospecting boreholes

◐ Diamond, drill rock cores taken

◖ Water flush—chip samples

◗ Hand auger in soil

Engineering site investigation boreholes

◓ Shell and auger

● Shell and auger with rotary core in rock

◑ Shell and auger with rotary in rock

◐ Rotary—rock roller

Research boreholes

⊙ Power auger—disturbed samples

⊗w Well, backfilled or inaccessible

⊕ Mine shaft, abandoned

▫ Trial pit

▪—▪ Sewer trench

⊔⊔⊔ Geophysics—constant separation resistivity traverse

12 Area of site investigation, with reference number

6 oc Opencast prospecting area, with reference number

Comment

A typical documentation map recording the location and nature of archival information from which engineering-geological and other maps of the area have been produced. The complete map, with National Grid co-ordinates, is linked to a punched-card data retrieval system; some of the data has been experimentally coded for computer storage and retrieval.

Part of the documentation map at a scale of 1: 10,000 being prepared for the engineering-geological survey of the Tyne and Wear Metropolitan County, northern England. The example presented here is from Dearman *et al*. 'Techniques of Engineering-Geological Mapping with Examples from Tyneside'. *The Engineering Geology of Reclamation and Redevelopment—Regional Meeting, Durham, Engineering Group, Geological Society*, p. 31–4, 1973.

Layout of descriptive memoir

6

Ideally, a comprehensive engineering geological map should be accompanied by a descriptive memoir containing the following information:

Contents

Introduction
    Purpose of the engineering geological map
    Geographical location of mapped area
        Topographical data
        Road, rail and other transportation routes
        Economic evaluation, and development prospects
    Previous investigations
    Methods used in the engineering-geological survey of the area
        Extent of the investigations
        Organizations which carried out the survey

Physical geography
    Climatic factors influencing the evaluation of engineering geological conditions
    Physiographic description
    Hydrography

Geological structure and development
    Pre-Quaternary
    Quaternary
    Present geodynamic processes

Geological characteristics of rocks and soils and their engineering geological properties
    Rocks
    Soils

Hydrogeological conditions
    Characteristics of individual aquifers
    Groundwater chemistry

Engineering-geological zoning
    Principles applied to the map area
    Characteristics of individual zoning units

Construction and other materials

Conclusions

Recommendations

Appendixes
    References
    Sources of archival and other material
    Tables of engineering geological properties

Index

The memoir would be illustrated by diagrams, graphs, tables and photographs.

The content of the memoir would, of course, be related to the purpose, content and scale of the map, and the layout of chapters given above is suggested as a basic guide from which suitable chapter headings may be selected.

Glossary

7.1 Introduction

In compiling this glossary, reference has been made to standard geological dictionaries written in English (Anon., 1962; Challinor, 1967; Gary *et al.*, 1972; Schieferdecker, 1959; Whitten and Brooks, 1972) and to the English version of the report of the Commission on Terminology, Symbols and Graphic Representation (Anon., 1970) of the International Society for Rock Mechanics. Purely geological, as distinct from engineering geological, terms used in the text are not defined if there is a satisfactory definition in either of two inexpensive and readily available English dictionaries of geological terms (Anon., 1962; Whitten and Brooks, 1972).

References are also given (7.3.2) to geological dictionaries written in other languages.

7.2 Definitions of terms used in the text

angle of internal friction   Angle of shear resistance; the angle (Ø) between the axis of normal stress and the tangent to the Mohr envelope at a point representing a given failure-stress condition for solid material.

archives   The place in which government or public records are kept (adj.: archival).

area   A taxonomic unit in engineering geological zoning delimited on the basis of the uniformity of individual regional geomorphological units.

attribute   A quality or property inherent in anything.

attrition value   *See* value, attrition.

classification   The formal arrangement into the groups of a hierarchy of taxonomic categories.

coefficient, permeability   The rate of flow of water through a unit cross-section under a unit hydraulic gradient.

coefficient, storage   The volume of water released from storage in each vertical column of the aquifer having a base of unit area when the water table or other piezometric surface declines by one unit of depth.

cohesion   Shear resistance at zero normal stress.

compaction   The packing together of soil particles with the expulsion of air only. It is accomplished by rolling, ramming or vibration, and results in a decrease in the air voids and an increase in the density of the soil.

component, engineering geological environment   The basic geological and geographical features which are of decisive significance for engineering geological mapping, namely the distribution and properties of rocks and soils, groundwater, characteristics of the relief, and present geodynamic processes.

compressibility   The decrease in volume per unit increase of pressure.

confined water   *See* water, confined.

consistency   Cohesive soils may be classified as stiff, firm or soft depending upon their consistency; the terms are indicative of the ease or difficulty with which the soil is excavated with a spade or moulded in the fingers. Consistency limits are the moisture contents at which a soil passes from the liquid to the plastic to the solid state.

creep   Time dependent deformation; the ability of rocks and other naturally occurring materials to be slowly, continuously and permanently deformed under loads over a long period of time.

deformability   *See* deformation.

deformation   A change in shape or size of a solid body.

density   Bulk density; the weight of a material, including the effect of voids whether filled with air or water, per unit volume.

discontinuity   An open, or potentially openable, structural plane such as a bed, joint, cleavage, fault.

district   A taxonomic unit in engineering geological zoning in which hydrogeological conditions and geodynamic phenomena are uniform.

durability   Power of resisting decay; resistance of a rock to weakening and disintegration when subjected to short term weathering processes. Slake durability, resistance to wetting and drying.

engineering geological component   *See* component, engineering geological environment.

engineering geological conditions   The dynamic geological system of the rocks and soils, water, geomorphological conditions and geodynamic processes at an individual site or area.

engineering geological map   *See* map, engineering geological.

engineering geological type   The mapping unit with the highest degree of physical homogeneity. It should be uniform in lithological character and physical state.

fall   Applied to mass movement (q.v.); downward and outward movement of slope-forming materials, in which the moving mass travels mostly through the air by free fall, leaping, bounding or rolling, with little or no interaction between one moving unit and another.

flow   Applied to mass movement (q.v.); downward and outward movement of slope-forming materials, in which the movement within the displaced mass is such that the form taken by the moving material, or the apparent distribution of velocities and displacements, resembles those of viscous fluids.

formation   The fundamental formal unit of lithostratigraphical (q.v.) classification; it is the only formal unit which is used for completely dividing the whole stratigraphical column all over the world into named units on the basis of lithostratigraphical character.

frost heave   The lifting of a surface by the internal action of frost.

geodynamic   Referring to those geological features of the environment resulting from geological processes active at the present time.

geohydrology   *See* hydrogeology.

geomorphology   That branch of both physiography and geology which deals with the form of the earth, the general configuration of its surface, and the changes that take place in the evolution of land forms.

geophysics   Geophysical method; geological exploration using the instruments and applying the methods of physics and engineering; exploration by observation of seismic or electrical phenomena or of the earth's gravitational or magnetic fields or thermal distribution.

groundwater   Subsurface water in the zone of saturation in the lithosphere.

group   A stratigraphical sequence of two or more contiguous formations having significant unifying lithological features in common.

homogeneity   Having the same property throughout.

hydrochemical   Referring to the chemical composition of natural waters.

hydrogeology   That part of hydrology which relates to the water in the lithosphere.

infiltration   The flow or movement of water through the soil surface into the ground.

in situ   In its natural position or place.

in situ test   *See* test, *in situ*.

iso-   Equal; a prefix, extensively used in conjunction with another word, to denote lines drawn on a map through points of equal value of the element displayed.

isobath   A line joining points of equal depth; for example, a line on a land surface all points of which are the same vertical distance above the upper or lower surface of an aquifer may be called an isobath of the specified surface.

isohypse   Contour of groundwater level or watertable.

isoline   Equal line; on an isoline map some variable feature is contoured.

isopachyte   Isopach; isopachous line; a line, on a map, drawn through points of equal thickness of a designated unit.

isopiestic   Isopiestic line; a contour of the piezometric surface of an aquifer.

isoseism   Isoseismic line; an imaginary line connecting all points on the surface of the earth where an earthquake shock is of the same intensity.

isotropic   Having the same properties in all directions.

landslide   Landslip; a portion of a hillside or sloping mass which has become loosened or detached and has slipped down.

land-use   Use of land by man.

legend   A brief discription of the symbols and patterns shown on a map or diagram.

lithogenetic   *See* lithogenesis.

lithogenesis   The origin and formation of rocks.

lithological complex   A mapping unit comprising a set of genetically related lithological types.

lithological suite   A mapping unit comprising many lithological complexes which are paragenetically related.

lithological type   A mapping unit which is homogeneous throughout in composition, texture and structure, but usually not uniform in physical state.

lithostratigraphy   Stratigraphy based only on the physical and petrographic features of rocks (adj.: lithostratigraphical).

map   A representation on a plane surface, at a specified scale, of the physical features of a part of the earth's surface or of any selected surface or subsurface data, by means of signs and symbols.

map, engineering geological   A type of geological map providing a generalized representation of all the components of a geological environment of significance in land-use planning, and in design, construction and maintenance as applied to civil engineering.

map, engineering geological, analytical   A map evaluating an individual component of the geological environment.

map, engineering geological, auxiliary   A map presenting factual data, for example a documentation map.

map, engineering geological, complementary   A map of basic geological and other non-engineering geological data.

map, engineering geological, comprehensive   A map of engineering geological conditions depicting all the principal components of the engineering geological environment; or a map of engineering geological zoning.

map, engineering geological, multi-purpose   A map providing information on many aspects of engineering geology for a variety of planning and engineering purposes.

map, engineering geological, special purpose   A map providing information on one specific aspect of engineering geology, or for one specific purpose.

map, interpretative   A general geological map interpreted in engineering geological terms.

map, large-scale   A map drawn at a scale of 1: 10,000 or greater.

map, medium-scale   A map drawn at a scale less than 1:10,000 and greater than 1:100,000.

map, small-scale   A map drawn at a scale of 1: 100,000 or less.

mass movement (syn.: mass wasting)   A general term for the dislodgement and downslope transport of rock and soil material under the direct application of gravitational body stresses.

mechanical property   *See* physical property.

modulus, deformation   The ratio of stress to corresponding strain during loading of a rock mass including elastic and inelastic behaviour.

permeability   The capacity of a rock or soil to conduct (transmit) liquid or gas. It is measured as the proportionality constant K between flow velocity v and hydraulic gradient I; $v = kI$. The unit of permeability is the darcy.

pH   The negative logarithm of the hydrogen ion activity.

photogrammetry   The science of obtaining reliable measurements from photographs.

physical property   A characteristic of rock or soil; a distinction may be made between a mechanical property which can be determined by a machine, e.g. uniaxial compressive strength, and a physical property which can be determined by the senses.

piezometric   Piezometric surface; an imaginary surface that everywhere coincides with the static level of the water in an aquifer; the surface to which the water from a given aquifer will rise under its full head.

point symbol   *See* symbol, point.

pressure   Force per unit area applied to the outside of a body.

pressure, uplift   The hydrostatic force of water exerted on or underneath a structure tending to cause a displacement of the structure.

region   A taxonomic unit in engineering geological zoning based on the uniformity of individual geotectonic structural elements.

resistivity method   A geophysical method of investigation in which the mean resistivity of the ground is measured and analysed.

rock   Strictly, any naturally formed aggregate or mass of mineral matter, whether or not coherent, constituting an essential and appreciable part of the earth's crust. In the engineering sense, hard, solid, and rigid deposits forming parts of the earth's crust. There are some naturally occurring materials which have properties intermediate between those of rocks and soils as defined here; they may be referred to as semi-solid rocks or soft rocks. *See also:* soil.

rock mass   Rock as it occurs *in situ*, including structural discontinuities and the effects of weathering.

rock material   Rock in the hand-specimen, generally excluding the structural discontinuities of the rock mass; rock material may be weathered and may contain micro-discontinuities, for instance incipient joints.

salinity   A measure of the quantity of total dissolved solids in water.

sample   A representative unit of a rock or soil material or mass.

sample, disturbed   A rock or soil sample which does not retain the characteristics of the *in situ* material.

sample, undisturbed   A rock or soil sample which retains, to a high degree, the characteristics of the *in situ* material.

seepage   The infiltration or percolation of water through rock or soil to or from the surface. The term seepage is usually restricted to the very slow movement of groundwater.

seepage force   The frictional drag of water flowing through voids or interstices in rock or soil causing an increase in the intergranular pressure, i.e. the hydraulic force per unit volume of rock or soil which results from the flow of water and which acts in the direction of flow.

seismic method   A geophysical method of investigation in which the travel times of refracted and reflected elastic waves through the ground are measured and analysed.

slide   The descent of a mass of soil or rock down a hill or mountain side.

slope movement   Mass movement on an inclined topographic surface.

soil   An aggregate of mineral grains that can be separated by such gentle means as agitation in water. The mineral grains may be uncemented or very weakly cemented as in sands and gravels, or may be bound together by weak forces, such as Van de Vaal's forces, as in silts and clays.

spring, suffusion   A natural upwelling of water carrying with it fine particles from unconsolidated materials. *See also:* suffosion.

storage coefficient   *See* coefficient, storage.

strength   Maximum stress which a material can resist without failing for any given type of loading.

strength, compressive   Maximum stress which a material can resist without failing in compression.

strength, shear   Maximum stress which a material can resist without failing by shear.

strength, tensile   Maximum stress which a material can resist without failing in tension.

structure (petrology)   The larger-scale interrelationships of textural features, generally seen or studied best in the outcrop rather than in hand specimen or thin section. Structure represents a discontinuity or major inhomogeneity, one of the larger morphological features of a rock mass (q.v.) such as jointing, bedding, cleavage, foliation. Not synonymous with texture (q.v.).

structure (structural geology)   The general disposition, attitude, arrangement or relative positions of rock masses of a region or area; the sum total of the structural features of an area, consequent upon such deformational processes as faulting, folding and igneous intrusion.

subsidence   Movement in which surface material is displaced vertically downwards with little or no horizontal component.

suffosion   misnomer for suffusion (q.v.).

suffusion   The washing out of fine particles from unconsolidated materials, in particular sands and gravels.

suffusion spring   *See* spring, suffusion.

surficial, superficial   Characteristic of, pertaining to, formed on, situated at, or occurring on the earth's surface; especially, consisting of unconsolidated residual, alluvial, or glacial deposits lying on the bedrock.

symbol, point   A generalized indication of the nature and location of a natural phenomenon.

water table   The plane which forms the upper surface of the zone of groundwater saturation.

taxonomy   The laws and principles of orderly classification (adj.: taxonomic).

terrain, terrane   The tract or region of ground immediately under observation.

test, *in situ*   A geotechnical test carried out in an excavation or borehole in which the rock or soil under test is in its natural position and state.

texture   The general physical appearance or character of a rock including the geometrical aspects of, and the mutual relations among, the component particles or crystals; e.g. the size, shape and arrangement of the constituent elements of a sedimentary rock, or the crystallinity, granularity and fabric of the constituent elements in an igneous rock.

uplift pressure   *See* pressure, uplift.

urbanization   A conversion of land to a city and its associated industrial and recreational facilities.

value, attrition   A value, obtained under standardized test conditions, of the resistance to wear of an aggregate stone sample.

water, confined   Groundwater that is under sufficient pressure to rise above the level at which it is encountered by a well or borehole.

water content   Moisture content; the percentage by weight of water contained in the pore space of a rock or soil with respect to the weight of the solid material.

water table   Groundwater surface, groundwater level; the level below which the rock and subsoil, to unknown depths, are saturated.

weathering That process of alteration of rocks and soils occurring under the direct influence of the hydrosphere and atmosphere.

zone A taxonomic unit in engineering geological zoning; based on lithological homogeneity and the structural arrangement of lithofacial complexes of rocks and soils.

7.3   References

Anon. 1962. *Dictionary of geological terms.* 2nd ed. New York, Dolphin Books.

Anon. 1970. *Terminology, symbols and graphic representation.* Commission of International Society of Rock Mechanics. 32 p. (Mimeo.)

BAULIG H. 1956. *Vocabulaire franco-anglo-allemand de géomorphologie.* Paris, Les Belles Lettres.

CHALLINOR, J. 1967. *A dictionary of geology.* 3rd ed. Cardiff, University of Wales Press.

COMECON. 1968. (*Terminological glossary on engineering geology.*) Moscow. 2 vols. 400 and 226 p. (In Russian with glossary terms in eight languages including English.)

GARY, M.; MCAFEE Jr. R.; WOLF, C. L. (eds). 1972. *Glossary of geology.* Washington, D.C., American Geological Institute.

MALJAVEEV, A. A. 1971. (*Glossary on hydrogeology and engineering geology.*) Moskva, Nedra. (In Russian.)

PLAISANCE, G.; CAILLEUX, A. 1958. *Dictionnaire des sols.* Paris, La Maison Rustique.

SCHIEFERDECKER, A. A. G. 1959. *Geological nomenclature.* Royal Geological and Mining Society of the Netherlands.

VISSER, A. D. 1965. *Dictionary of soil mechanics in four languages: English/American, French, Dutch and German.* Amsterdam, Elsevier; Paris, Dunod.

WHITTEN, D. G. A.; BROOKS, J. R. V. 1972. *The Penguin dictionary of geology.* Harmondsworth, Penguin Books Ltd. (Penguin reference books.)

Select bibliography

8

8.1 Introduction

The bibliography sets out a selection of papers on engineering-geological maps and mapping and some examples of published engineering geological maps. Some of the papers contain examples of engineering-geological maps, but neither these nor the published maps follow the recommendations set out in this guidebook in their entirety.

8.2 Engineering geological mapping

ARNOULD, M.; VANTROYS, M. 1970. Essai de cartographie géotechnique automatique sur la ville nouvelle d'Evry (Région parisienne). *International Association of Engineering Geology. First International Congress*, vol. 2, p. 1069–80.

BACHMANN, G.; GRÖWE, H.; HELMERICH, K.; REUTER, F.; THOMAS, A. 1967. Instruktion für die Anfertigung einheitlicher ingenieurgeologischer Grundkarten. *Zentr. Geol. Inst. Abh.*, vol. 9.

BLANC, R. P.; CLEVELAND, G. B. 1968. *Natural slope stability as related to geology, San Clemente area, Orange and San Diego counties, California*. 19 p. (Calif. Div. Mines and Geology special report, 98.)

BREDDIN, H.; VOIGHT, R. 1967. Eine Baugrundplanungskarte für Landkreis Kempen-Krefeld/Niederrhein [A building ground planning map for the district Kempen-Krefeld/Niederrhein]. *Geologische Mitteilungen*, vol. 7, p. 239–50.

BURTON, A. N. 1970. The influence of tectonics on the geotechnical properties of Calabrian rocks and the mapping of slope instability using aerial photographs. *Q. J. Engng Geol.*, vol. 2, p. 237–54.

CHURINOV, M. V., TSYPINA, I. M.; LAZAREVA, V. P. 1962. Principles and methods of compilation of universal areal maps of the U.S.S.R. at scales 1:1,500,000 to 1:2,500,000. *Soviet Geology*, no. 11, p. 112–24. (In Russian.)

—. 1970. The principles of compiling the engineering-geological map of the U.S.S.R. territory on the scale of 1:2,500,000. *International Association of Engineering Geology. First International Congress*, vol. 2, p. 850–60.

DAGNEUX, J. P.; LEMOINE, Y. 1970. Cartographie géotechnique en milieu côtier vaseux par sismique réfraction et pénétromètre. *International Association of Engineering Geology. First International Congress*, vol. 2, p. 895–903.

DEARMAN, W. R. *et al.* 1972. The preparation of maps and plans in terms of engineering geology. *Q. J. Engng Geol.*, vol. 5, p. 293–381.

DEARMAN, W. R., MONEY, M. S.; COFFEY, J. R.; SCOTT, P.; WHEELER, M. 1973. Techniques of engineering-geological mapping with examples from Tyneside. *The engineering geology of reclamation and redevelopment—Regional Meeting, Durham, Engineering Group, Geological Society*, p. 31–4.

FOOKES, P. G. 1969. Geotechnical mapping of soils and sedimentary rocks for engineering purposes with examples of practice from the Mangla Dam project. *Géotechnique*, vol. 19, p. 52–74.

GAZEL, J.; PETER, A. 1969. Essais de cartographie géotechnique. *Annales des Mines*, Dec., p. 41–60.

GOLODOVSKAYA, G. A.; DEMIDYUK; L. M. 1970. The problem of the engineering and geological mapping of deposits of mineral resources in the area of eternal frost. *International Association of Engineering Geology. First International Congress*, vol. 2, p. 1049–68.

GRABAU, W. E. 1968. An integrated system for exploiting quantitative terrain data for engineering purposes. In: G. A. Stewart (ed.), *Land evaluation. CSIRO Symposium*, p. 211–20. Canberra, Australia, Macmillan.

GRANT, K. 1968*a*. A terrain evaluation system for engineering. *Commonwealth Sci. Indus. Research Organization Australia, Div. Soil Mech., Tech. Paper 2*, p. 27.

—. 1968*b*. Terrain classification for engineering purposes of the Rolling Downs Province. *Commonwealth Sci. Indus. Research Organization Australia, Div. Soil Mech., Tech. Paper 3*, p. 385.

HUMBERT, M. *et al.*, 1971. Mémoire explicatif de la carte géotechnique du Tanger au 1:25000. Contribution à la connaissance du Tangérois. *Notes et M. Serv. géol. Maroc.*, vol. 222, p. 190.

JANJIC, M. 1962. Engineering-geological maps. *Vesnik Zavod za Geoloska i Geofizicka Istrazivanja* (Bull. Inst. Geophys. Res), ser. B, no. 2, p. 17–31.

LOZINSKA-STEPIEN, H.; STOCKLAK, J. 1970. Metodyka sporzadzania map ingyniersko-geologiscznych w skali 1:5000 i wiekszych [Mapping methods for geological engineering 1:5000 and larger scale maps], *Geological and engineering investigation in Poland*, vol. 5, *Instytut Geologiczny Bull.*, vol. 231, p. 75–112. (In Polish with English and Russian summaries.)

LUNG, R., PROCTOR, R. (eds.). 1966. *Engineering geology in southern California*. 389 p. (Assoc. Engineering Geologists Los Angeles Section special pub.)

MATULA, M. 1969. *Regional engineering geology of Czechoslovak Carpathians*. Bratislava, Publishing House Slovak Academy of Sciences.

—.1971. Engineering geologic mapping and evaluation in urban planning. In: D. R. Nichols and C. C. Campbell (eds.), *Environmental planning and geology*, p. 144–53. (Geological Survey, United States Dept. of the Interior and the Office of Research and Technology, United States Dept. of Housing and Urban Development.)

NICHOLS, D. R.; CAMPBELL, C. C. (eds.). 1971. *Environmental planning and geology*. 204 p. (Geological Survey, United States Dept. of the Interior and the Office of Research and Technology, United States Dept. of Housing and Urban Development.)

NILSEN, T.H. 1971. *Preliminary photointerpretation map of landslide and other surficial deposits of the Mount Diablo Area, Contra Costa and Alamenda Counties, California.* (United States Geol. Survey Miscellaneous Field Studies Map MF-310.)

POPOV, I. V.; KATS, R. S.; KORIKOVSKAIA, A. K.; LAZAREVA, V. P. 1950. *Metodika sostavlenia inzhenerno-geologischesikh kart* [The techniques of compiling engineering geological maps]. Moskva, Gosgeolizdat.

SANEJOUAND, R. 1972. *La cartographie géotechnique en France.* Ministère de l'Equipement et du Logement, p. 96.

8.3 Published engineering geological maps

BRABB, E. E.; PAMPEYAN, E. H.; BONILLA, M. G. 1972. *Landslide susceptibility in San Mateo County, California.* (United States Geol. Survey Miscellaneous Field Studies Map MF-310.)

BROWN Jr., R. D. 1972. *Active faults, probable active faults and associated fracture zones, San Mateo County, California.* (United States Geol. Survey Miscellaneous Field Studies Map MF-355.)

BRYANT, B. 1972. *Folio of the Aspen Quadrangle, Colorado.* (United States Geol. Survey Miscellaneous Geologic Investigations Map I-785-A through G.)

CHRISTIANSEN, E. A. (ed.) 1970. *Physical environment of Saskatoon Canada,* p. 68. (Saskatchewan Research Council, Nat. Research Council Canada Pub. 11378.)

DEBAILLE, G.; GHISTE, S. 1969. *Carte géotechnique de la région de Mons.* Mons, Institut Reine Astrid. 44 p., 4 maps.

FISHER, W. L.; MCGOWEN, J. H.; BROWN Jr. L. F.; GROAT, C. G. 1972. *Environmental geologic atlas of the Texas Coastal Zone Galveston-Houston Area.* Bureau of Economic Geology, The University of Texas at Austin. 91 p.

FÜLÖP, J. (ed.). 1969. *Engineering-geological map series (scale 1: 10,000) of the environs of Lake Balaton.* Tihany, Budapest, Hungarian Geological Institute.

RADBRUCH, D. H. 1969. *Areal and engineering geology of the Oakland East Quadrangle.* (United States Geol. Survey Geol. Quad. Map GQ-769.)

RONAI, A. 1969. *The geological atlas of the Great Hungarian Plain. Scale 1: 100,000.* Budapest, Hungarian Geological Institute.

UNITED STATES GEOLOGICAL SURVEY. 1967. *Engineering geology of the North-east Corridor, Washington D.C., to Boston, Massachusetts.* (United States Geol. Survey Miscellaneous Geologic Investigations Map I-514-A through C (scale 1: 250,000).

WILLIAMS, P. L. *et al.* 1971–73. *Folio of the Salina Quadrangle Utah.* (United States Geol. Survey Miscellaneous Geologic Investigations Map I-591-A through N.)

Acknowledgements

9

Members of the commission would wish to record their indebtedness to those who have freely made available the original maps on which the illustrations for Chapter 5 have been based. A great number of original drawings had to be made in order to produce the map examples in colour. Drawings for maps 5.2.2.1, 5.2.2.2 and 5.2.2.3 were made in the Department of Engineering Geology and Hydrogeology, Comenius University, Bratislava (Czechoslovakia); the legends for these maps and the drawings for all the other maps were undertaken by Eric Lawson, Department of Geology, University of Newcastle upon Tyne (England). Miss A. Thwaites and Mrs S. Gaynor of the same department typed the many early versions of the text and the final manuscript. Without this very considerable help, which is gratefully acknowledged, this first major achievement of the IAEG Commission on Engineering Geological Maps could not have been brought to fruition.

[A.38] SC.74/XVII.15/A